Dedications

To Moana and Jack

my wife Dian

and

The 13th Corps
1st Battalion Prince of Wales's
North Staffordshire Rifle
Volunteer Regiment

13th CORPS
A SHORT HISTORY

1st BATTALION
NORTH STAFFORDSHIRE
RIFLE VOLUNTEERS
(1859 – 1908)

(KIDSGROVE & DISTRICT 1859 – 1896)

(GOLDENHILL & DISTRICT 1896 – 1908)

Alan J. Jones

Canberra, 2021

First published in Australia in 2022

ISBN: 978-0-6452924-4-2

Introduction and Acknowledgements

It was David Wood the Convenor of the Goldenhill Memories Group that brought my attention to the Kelly's Directory of Staffordshire 1896, which he accessed through the University of Leicester digital collection and discovered the existence in Goldenhill of the 13th Corps or Company-F of the 1st Battalion Prince of Wales's, North Staffordshire Rifle Volunteer Regiment under the command of Captain W. H. Grindley, the famous pottery manufacturer of Newfields and Tunstall.

This discovery raised a whole series of issues in the annals of the history of the small Goldenhill village. Why would a small village have an army corps and what would this mean in a growing industrial community?

Little did we guess at the time of this discovery that it would also reveal two significant and quite massive movements in both the Military and Social History of Great Britain as a whole. It rapidly became evident that we needed not just to look at 1896, but to also step back even further in history to investigate the drivers of these changes.

Interestingly, my starting point was triggered mostly by the movie "Peterloo" which readers may recall is set at the end of the era of Napoleon Bonaparte and the Battle of Waterloo. Peterloo is of course a play on the word Waterloo, but more importantly, the story-plot of the movie really outlines the inequalities that existed in British society at the beginning of the 19th Century, not just in terms of differences in status and wealth but also in the differences in human rights, the absence of the right to vote, and thus the lack of Parliamentary representation for the rapidly growing urban populations. The movie brings into play the Chartist movement as a driver for change, but far worse it shows the blatant disregard of the "establishment" on the topic of people's rights, exemplified by bringing in the Military to quell a so-called "riot", with the resultant loss of life described as the "Peterloo Massacre" of August 1819.

Some years later (1842) the lesser known but more violent "Potteries Riots" occurred, but with only one death inflicted by Military intervention though the toll on the demonstrators through use of long-term deportations to Australia was inflicted on 54 individuals (21 were potters and 19 miners). The post-Napoleonic period for the Borough (Stoke-on-Trent) in which Goldenhill resides is elegantly described in John Ward's Book: *Borough of Stoke-on-Trent*, published in 1843.

I want to thank several members of the Goldenhill Memories Group for assisting me in this project with their memories and intimate connections with the old village. Most notably, David Wood, David Salmon, Alan Goodwin and Bill Rocky; and many other GMG members for their inciteful comments as I progressed in this endeavour.

I am particularly indebted to the diligent staff of some of the archival agencies of Staffordshire, notably the Staffordshire Record Office, the William Salt Library and the City of Stoke-on-Trent Archives. Very specifically I want to thank Jonathan Goodwin, City of Stoke-on-Trent, for his unearthing the Birmingham Newspaper article covering the conception of the 13th Corps for Kidsgrove and District, and his insight into connecting the words "Drill Hall" to a Sunday School at Christ Church, Goldenhill. Finally, I am indebted to the two wives of my life, the first born in Goldenhill and attending the Christ Church premises as a youngster with her memories of that site; and the second for her incredible patience in bearing with my attention to follow this project through to a conclusion, and her contributions in reading and editing the text. The photograph of "Rodgers Street" is Reproduced Courtesy of Stoke-on-Trent City Archives. The main body of research was assisted through digital access to the British Newspaper Archives (britishnnewspaperarchive.com.uk); The Kelly's Directories of Staffordshire that are accessible; and the Army Lists/Monthly Army Lists. I also want to particularly thank Mr Peter Collins, Museum Director of the Staffordshire Regiment Museum for supplying me access to the photographs of the 1st Battalion used in this text. On the front cover is a photograph Group at Camp, 1884. Courtesy of the Trustees of the Staffordshire Regiment Museum. Incidentally no one seems to know the exact colour of the uniform worn by the 1st Battalion. An early painting indicates a light/brown/khaki, but the Museum photographs suggest something darker, like a bottle green or blue.

I am particularly grateful for the existence of the British Newspaper Archive, which I feel sure holds many more treasures than I have been able to explore to date.

I also want to thank and Philip Leese (Historian) and Mick Maloney of Kidsgrove for their generosity in keeping me informed on some matters generally not readily accessible in the public domain, especially for someone working on this topic outside the UK.

Alan J. Jones 2021

TABLE OF CONTENTS

LIST OF TABLES AND ILLUSTRATIONS

ABBREVIATIONS USED IN THE TEXT

Ref Reference

Loc Indicating location number in a specific Kindle Book

SSCGA Staffordshire Sentinel and Commercial and General Advertiser

SS Staffordshire Sentinel

RVC Rifle Volunteer Corps

VD Volunteer Decoration

Chapter 1. A Growing Military Need

It appears historically for time immemorial that the British Isles have been under threat from foreign invasions. Notably, from the merging of Britain into the Roman Empire, the raids of the Anglo-Saxons, the raids of the Viking hordes, and then the Normans. It was not until the times of Queen Elizabeth the First and the defeat of the Spanish Armada that Britain's Naval strength was acknowledged as the most superior in the then known world. European revolutions followed in the course of history, particularly in France, which placed a different perspective on militarism all over Europe, and the evolution of Napoleon Bonaparte who had aspirations of conquering the entire European world, including Russia. When Napoleon threatened invasion of Britain by assembling his "Army of England" on the French shores across the Chanel, the then King and Parliament called for Volunteer Corps to be formed in various parts of the country.

Here, I must call on the remarkable history of the Borough written by John Ward in 1843 (Ref 1). He notes, and I paraphrase, that there had been many earlier "calls to arms" that had resulted in the formation of local Militia groups, but this new call was on much broader scale and relied on the expression of loyalty by each individual and each district to "King, Queen, Government and Country" with the aim of providing defence of the homeland and guaranteed peace in their own neighbourhoods. The local response (in the Borough of Stoke-on-Trent) to this call to arms, was the rapid formation of a troop of Cavalry, under Captain Commandant Sir John Edenson Heathcote of Longton. The troop consisted of about seventy young men from various parts of the Borough, mounted and equipped at their own expense.

There followed several corps of Volunteers accepted by the Government as suitable for military training on the provision that they would be armed and equipped at their own individual expense, or by local subscription. Tunstall created three companies of eighty rank-and-file each commanded by Major John Davenport (at that time to become a local representative in Parliament). Burslem generated a corps of Volunteer Infantry of about seventy men, commanded by Captain William Sneyd of Bradwell. Similarly, the Stoke, Penkhull, and Fenton volunteers formed corps of four companies, commanded by Lieutenant-colonel Daniel Whalley; and the Lane End volunteers formed another corps of four companies, commanded by Major William Turner. Hanley and Shelton volunteers mustered six companies, commanded by Lieutenant-colonel James Whitehead. Thus, a local force of more than 1,400 men, including officers, was generated. Such numbers swelled throughout the nation.

These corps were disbanded when the "time of peace" arrived (Ref1, p56). The Volunteers corps had been called out on only a few occasions, for example, "to quell the occasional tumultuous proceedings in the years of scarcity during 1800 and 1801", when the populace proceeded to break into the corn and flour shops, and to vend the produce at their own prices. It was said that the humanity of Colonel Walter Sneyd showed through on the most trying of these occasions, and this met with universal approbation at the time. (Sneyd happened to be at Keele, instead of Windsor, where he was usually stationed in attendance on King George III).

As an anecdotal diversion I note that such calls on Volunteers occurred in other parts of Europe. I draw my own experience in the 20th Century – albeit, that is, jumping forward in time some 200 years later I had the pleasure of being appointed as a Senior Diplomat for Australia to the Federal Republic of Germany. The Australian Embassy was based in the then Capital of Germany, Bonn. Within weeks of arrival in Bonn, I received an invitation to join a German Volunteer Military group. I was given a senior Officer rank of Rittmeister (Riding Master) in the Ehrengarde (Honour Guard) Corps of Köln (Cologne, just 13 kilometres north of Bonn). They issued me with a minimal uniform (a hat and numerous medals) in what proved to be "Australia's colours of Green and Gold". This regiment was one of many that formed in the Napoleonic era covering the entire length of the River Rhine. Today the Ehrengarde exists as a community voluntary organisation raising money for all manner of good causes through their "Ump-pa-pa Band" of Brass and Timpani; its Dance Troup – providing entertainment at large municipal functions; but more importantly its involvement in the coordination of a series of massive parades/street festivals called "Carnival" that goes on over two weeks in late winter/early spring. Many of the guests at the Ehrengarde functions were real Military officers from every country in the world, but most memorably the UK and Russia. I often felt outranked by some of the Generals, but it was all good fun.

My wife and I at an Ehrengarde Corps Celebration in 1992

Following Napoleon's defeat at Waterloo the French remained uncomfortable with their standing in the world. By 1850 they had grown their army to about 400,000 regulars, supported by 2 million "Garde National". In contrast, the British Army amounted to 120.000 regulars, with barely 37,000 stationed in Britain. By 1858 the total number of British Regulars and existing auxiliary troops totalled 153,285 but only about a third of these were considered to be fit enough for active duty (Ref 2, Loc 511). Recall that this was the time of major British Imperialism having established elements of the British Empire all over the world.

In writing this story as a local piece for Goldenhill I drew attention to the Map of the World, which in my childhood was to be found in every school room in the land. Readers may recall the flat Mercator Projections of the globe. When cartographers were asked to flatten the Earth for such a map projection, they noted the distortions that creep in, in terms of shape, distance, direction, or land area creep. With Africa in the centre of the Map, you may recall all the "pink areas" that represented the British Empire, which looked quite massive at the outer extremities, particularly the sizes of Canada (top left) and Australia (bottom right) – "clever political stuff in its day". Add to that a good deal of Africa, all of the Indian continent, and parts of Asia comprised the British Empire package. Children were taught to acknowledge that our "Country was Great!". But, of course. to be "Great" our country had to have Military forces in each and every one of those pink areas, to quell uprisings or even fight Wars as, for example, in parts of India; the Zulu wars; the Boer War and so on. Britain was at its height in Military demand to meet its' commitments overseas.

At this time Napoleon the Third had achieved a Coup d'état in taking over the Government of France, and he maneuvered the Government into enlarging their military forces. In addition, Napoleon III also embarked on programs to increase French Naval capability, even to their becoming the first nation to launch an "iron-hulled ship", conceptually at least, driven by the "new" steam engine.

The fear of an "invasion" became high on the political agenda once again by 1858 and following long debates about the pros and cons of Volunteer forces Britain saw a need to establish new Volunteer home-based forces yet again. A young Queen Victoria was led by her Parliamentary advisors to call for the formation of a Volunteer Army throughout Britain. The motivation for this was not just external military invasion, but also to help supplement our regular forces who were somewhat overcommitted in all the theatres they occupied within the Empire; and in my view to keep the thoughts of the working classes otherwise occupied, since there were increasing signs of the desire for parliamentary representation, and militant unionism was on the horizon.

The response to forming Volunteer Corps in the Potteries, again an expression of loyalty to Queen and Country, was quite massive. The Corps that were created by 1859, with the permission of the Queen were called the Rifle Volunteer Movement, comprising Corps, simply described as Rifle Volunteer Corps (RVC). The 2nd Staffordshire RVC was raised in Longton on 30 September 1859, and was quickly followed by others listed below (dates of formation in parentheses). By May 1860, there were enough company-sized (70 -100 men) RVCs in the Stoke area to be formed into the 1st Staffordshire Administrative Battalion, comprising (Ref 3):

- HQ at Stoke-on-Trent
- A Company at Longton (original 2nd Staffs RVC formed 30 September 1859)
- B Company at Hanley (3rd Staffs RVC formed 27 September 1859)
- C Company at Burslem (6th Staffs RVC formed 28 December 1859)
- D Company at Tunstall (9th Staffs RVC formed 4 January 1860)
- E Company at Stoke-on-Trent (10th Staffs RVC formed 19 January 1860), with an affiliated Cadet Corps from 1875 to 1884.

- F Company at Kidsgrove (13th Staffs RVC formed 26 February 1860)
- G and H Companies at Newcastle-under-Lyme (16th Staffs RVC formed at the Old Barracks at Newcastle-under-Lyme on 24 February 1860)
- J Company at Leek (28th Staffs RVC formed 26 April 1860)
- K Company at Hanley (36th Staffs RVC formed 18 June 1860)
- L Company at Stone (40th Staffs RVC formed 1 December 1860)
- M Company was formed later at Trentham.

Rumour has it that the Tunstall Company had designs on being enlarged by incorporating the Kidsgrove District into their ranks, but that was not to be. The "power" of the landed-gentry came into play here. In reality the Kidsgrove and District Company-F was navigated into its place in history by the then well-known Williamson family, who at this time were living at Ramsdell Hall near Congleton in Cheshire, but who were multiple owners of most of the mineral resources (coal and iron ore) being mined throughout North Staffordshire, especially around Goldenhill and Kidsgrove. While the term "F-Company at Kidsgrove" might appear quite correct this Company was actually formed from a consolidation of input from the entire Kidsgrove District.

A Newspaper Report (Birmingham Daily Post, 1 December 1859) provides the background. A public meeting was interestingly called by the Church Wardens of KIdsgrove, Mow Cop, Newchapel, Goldenhill and Talk-o'-t'-hill and was held at the school room in Kidsgrove to discuss the best means of forming and equipping a volunteer rifle corps for the named districts. Attendance at this meeting was described as "numerous and influential", including Hugh H. Williamson, Edward Williamson, the Reverend Wade (Kidsgrove), the Reverend Forshaw (Newchapel), the Reverend M'Hutchin (Talk), William Rigby, T Brindley, T Sherrat, Messer's Gater (father and son), T Hesketh, and J. B. Davenport.

The Reverend Wade Presided at the meeting, commencing with a stirring speech about the need for self-defence in case England should be assailed. He stated that: "England was not alarmed, it was below our dignity to be alarmed". He pointed to the growth in French militarism. This was a time when England expected everyman would do his duty. H. H. Williamson (HHW) followed with a proposal for the formation of a

Volunteer Rifle Corps within the united parishes. Mr John Gator Snr., seconded. Edward Williamson proposed a second resolution "That a deputation be appointed to wait upon the inhabitants of the district, to solicit their cooperation, for the purposes of providing funds to meet the necessary expenses". This was seconded by Mr Hesketh. By the time the meeting was concluded the meeting had raised £205, including £100 from HHW, and 34 volunteers had enrolled.

The machinations of the proposed formation had a few loops to go through (all completed by 26 February 1860), but this was the beginning of Company-F that about 30 years later became the Company-F operating out of Goldenhill.

Chapter 2. Growth in the Local Military Presence

There exists a trail of newspaper articles coupled with the Olympic-style repetition on a 4-year cycle of the Kelly's Directories, and the almost monotonously monthly publications by the War Office initially called the "Army Lists", but later "Monthly Army Lists". These latter titles provide lists of the Officers involved in every sector of the British Army since 1815, including promotions, achievements and retirements. Unfortunately, the early editions are now highly fragmented, but to our benefit we have access to these reports, at least annually from 1859 to 1914, though changes in format limit their viability in the absence of other information, for example, from the Kelly's Directories and newspaper articles, unless the research being conducted is at the Regimental Officer level.

The study of the collective data provides a more definitive picture of Goldenhill's involvement in the military establishment of the 19th Century, notably in the Rifle Volunteer Corps (RVC). It should be remembered, however, that underlying such an analysis are the responses of society to rapid industrialisation, including the massive growth in working-class populations, and the expression of local solidarity and aspirations of the newly formed communities. Among other things, this period in the 19th Century frankly witnessed the wanton destruction of many local environments, but also the creation of previously unheralded entertainments like rifle clubs, and brass bands, accompanied by numerous local and national competitions. In many ways a new way of life for most of the people involved

Concentrating on the formation of the Kidsgrove and District Rifle Volunteer Corps. Following the public meeting of November 1869 described above, the Staffordshire Advertiser of 14 January 1860 relates a "story" that the original proposal for a corps of 60 men could at this point be increased to 100 men, with nominated officers: Mr William S Williamson as Captain, Mr Edward Williamson as Lieutenant, and Mr Thomas Brindley as Ensign. A shooting range at Harecastle Hill was nominated, and a sum of £490 had been raised, with annual subscriptions of £25 also guaranteed. In addition, the preferred services of the Kidsgrove local band had been accepted, which effectively provided another 15 recruits to Company-F.

It is of some interest to compare the rapid success in forming the KIdsgrove and District Rifle Volunteer Corps with the failure to create a Rifle Corps movement at Rochester in Kent, described in formal letters published in the Press (Reynolds's Newspaper, 27 November 1859).

By that time the KIdsgrove Corps was getting off the ground, the Tunstall Corps was fully operational and in a letter to their assembly (SSCGA: 28 January 1860) Mr Hugh H. Williamson (the proposer of the Kidsgrove Corps) noted that he was himself a Volunteer in the 1798 Burslem Corps. He probably was, but the Burslem Corps was short-lived when John Wood (son of the renowned potter Enoch Wood) presumed Commander of the corps was requested to discontinue the Burslem activities in October 1803. (We note later that HHW became commander of the Longton Corp at the retirement of John Davenport the founding commander of that corps).

FAILURE OF A RIFLE CORPS MOVEMENT.

The population of Rochester and the adjoining towns, say 35,000, were invited to share in defending the coast of Kent, and on the 7th September a public meeting was held in the Guildhall, to inaugurate the formation of a voluntary artillery corps. A small minority ventured to suggest that if the coast were still defenceless it might be a duty to glance at the largely-increased votes recently made for these very defences, and to ask the Government what had been done with the money. The sympathies of perhaps a dozen led them to vote this way, but the bulk of the assembly enthusiastically carried the resolution. The result is seen in the Mayor's letter as under, addressed to the Lord-Lieutenant of the county:—

"Rochester, Nov. 2, 1859.

"My Lord,—I acquainted your lordship that a public meeting was held on the 7th September, in the Guildhall of this city, and that it was then resolved to raise a volunteer corps of artillery out of the inhabitants of these towns. It is with great regret I now inform your lordship that the attempt has entirely failed.

"I have every reason to believe that donations and subscriptions fully equal to the formation and maintenance of a corps of artillery would have been forthcoming; but with shame I say it, that the committee appointed on the 7th of September last have not received the offer of a single volunteer.

"In conclusion, I beg to assure your lordship that the apathy or want of public spirit thus exhibited cannot give your lordship more annoyance and disgust than it does your lordship's very obedient servant,

"WILLIAM MANCLARK,
"Mayor of Rochester.

"The Lord Viscount Sidney, Lord-Lieutenant of the County of Kent."

By 11 February 1860 (SSCGA), her Majesty Queen Victoria had accepted the services of the Kidsgrove RVC – now nominated as the 13th Corps in the County of Staffordshire. This more or less coincided with the formal recognition of the Volunteer Movement by the Queen (SSCGA, 10 March 1860).

On 28 April 1860 (SSCGA) Lord Hatherton, the Lord-Lieutenant of Staffordshire submitted a plan to her Majesty for the formation of 31 companies of Rifle Volunteers in Staffordshire divided into 4 Battalions. The 1st Battalion comprising Longton, Hanley, Burslem, Tunstall, Stoke, Kidsgrove, and Newcastle Companies. This was just in North Staffordshire, but the Volunteer Rifle Movement took off in almost every corner of Britain. Incidentally, the Hugh Henshall Williamson discussed is the second-generation Williamson with the initials HHW, who is most readily identified by being the holder of the Greenway Bank Estate, where he passed away childless in 1867. He was also one of the most extensive coal masters in North Staffordshire, and is described as "warmly attached" to the Brown Edge and Norton districts through his support of church, parsonages and schools (SSCG, 7 December 1867).

It is worth noting at the national level that by November/December 1859 there were approximately 700 recruits reporting for duty daily in RVCs around Britain. There followed an average monthly increase over the following two years to May 1861 of 7,000 recruits. In 1859 there were 133 Rifle Corps formed, increasing to a total of 578 in 1860, and a further 36 in 1861 (Ref 2, Loc 796 and Loc 810).

Further, the information on the founding of the 13th Corps, Kidsgrove and District F Company demonstrates another underlying principle of RVCs, namely that each corps was intended to be self-sufficient, be it through self-funding of members of the corps, donations from major sponsors (local landowners or factory owners), or annual subscriptions made by participating officers and others. It was estimated that in 1860 the cost to the tax payers of assembling the total RVC movement was of the order of £3,000 (Ref 2, Loc 3150). Within quite a short time the realities of self-sufficiency proved quite unrealistic, especially when it came to finding funds to pay for disposables (e.g., ammunition), and sometimes replacing damaged or buying new equipment. Requests to Government initially led to "Capitation Grants" to cover the costs of munitions, an annual allowance of twenty shillings per Company, followed by calls for support for Band instruments, new armaments as technological advances occurred, and even new uniforms. By 1897

Government grants to support the RVC had reached £627,200 (Ref 2, Loc 3149). Such sums led to the development of increased Government control over time.

Officers of the 1st Battalion North Staffordshire Regiment Rifle Volunteers at camp in full dress uniform in 1884. Courtesy of the Trustees of the Staffordshire Regiment Museum.

In a recent paper, Philip Talbot (Ref 10), provides an abstract of the accounts of the 1st Volunteer Battalion North Staffordshire Regiment for the eighteen months 1 April 1884 to 30 September 1885, which serves to demonstrate many of the costs of operating a Volunteer Battalion, and at the time some of the main sources of income. Notable expenses relate to clothing, band expenses and wages, prizes, and camp expenses, most of the latter would have been unforeseen at the time of establishment.

ACCOUNTING IN THE LATE VICTORIAN VOLUNTEER ARMY 117

TABLE 1

An Abstract of the Accounts of the 1st Volunteer Battalion North Staffordshire Regiment
[for eighteen months from 1 April 1884 to 30 September 1885]

	£ s. d.		£ s. d.
Subscriptions	477.01.00	Ammunition	80.02.10
Hon. Members' Subs	47.05.00	Clothing	202.16.06
Ammunition Receipts	79.05.11	Sergeant Instructor	57.10.00
Band A/C	120.00.00	Band Expenses	67.10.05
Govt Grants:		Band Wages	168.16.04
Company A	154.10.00	Shooting Butts	60.00.06
B	160.00.00	Camp Expenses	755.18.05
H	154.10.00	Printing and Stationery	4.02.09
Camp Allowance:		Rent of Armoury	30.00.00
Company A	135.02.07	Coal and gas	57.06.00
B	136.06.07	Asst Sec. salary	12.12.00
H	129.02.07	Postages	17.11
Sale of old uniforms	3.17.00	*Volunteer Service Gazette*	19.00
New Band Clothing	1.08.11	Recruit Prizes	11.00.00
		Hire of skating rink for drills	26.12.00
		Band prizes	24.00.00
		Drill Instructor	1.10.00
		Prizes	1.12.06
		Shooting Prizes	210.00.00
		Sundries	6.04.06
Bal. due to bank	173.06.01		
	1779.11.08		1779.11.08

Chapter 3. Early Activities of the F-Company

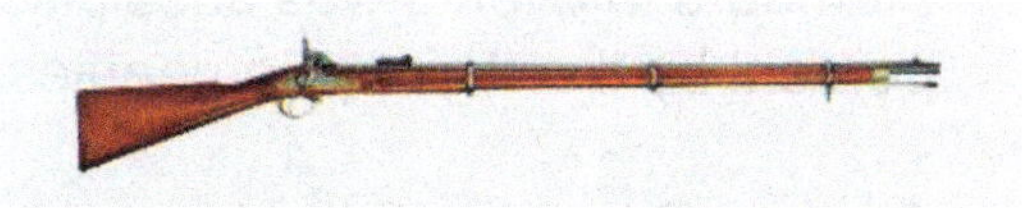

It is interesting that in its early days of formation the KIdsgrove and District Company received some bad Press due to their absence at several early Battalion drill parades held at Trentham Gardens and Fenton Manor in July/August 1860 (SSCGA, 28 July and 18 August 1860). The 13th Company was just not ready, that early on.

One of the first public appearances of Company-F, the Kidsgrove and District Rifle Volunteers took place on 19 March 1860 (SSCGA), when they were reported as assembling at the Armoury in Kidsgrove, headed by their Band to march to the railway station. They proceeded by train to Congleton where they were met by Captain Williamson. The report notes that after going through a few drill evolutions they marched through Congleton and were greeted at Mortlake (we are unable to ascertain if this is Mortlake the village, or Mortlake House, a larger but quite modest house in the district) by Mrs Williamson and other neighbouring gentry including Sir Charles Shackerley (a Commander of the Cheshire RVCs). Incidentally, travel by rail of uniformed Corps members to events was subsidised by fifty percent by the railway companies.

A proud Company-F then marched through Astbury to the Bleeding Wolf, Hall Green for Dinner followed by numerous toasts. By 11 pm the Company had returned to their HQ and were dismissed to their homes. The Kidsgrove and District Volunteers were to return to Astbury on several future occasions, but another very notable event occurred 20 years later, at a divine service held in honour of Lieutenant Hugh W Williamson on his retirement from the Company (SSCGA, 11 September 1880). Other attendees on this occasion included ten sergeants, 109 rank-and-file members of the corps, and representative officers from the Tunstall Company.

A similar Parade and Camp, but of the entire 1st Battalion at Whittington Heath in Lichfield in 1888, Courtesy of the Trustees of the Staffordshire Regiment Museum

On another prestigious occasion the members of the 13th corps assembled at their parade ground in Kidsgrove and marched to Ramsdell Hall (although not recorded we suspect using the train service to Congleton) on Monday the 29th September 1860 (SSCGA) to be presented with a silver bugle, "the munificent gift of a kind friend". Many people assembled to witness the presentation, and among those present were Major Williamson (HHW), Adjutant Buchanan, Captain Midgley, R Williamson jnr., P Williams, G.W. Reade of the South Cheshire Regiment, Ed J Solly (the Mayor of Congleton), Arthur L Solly, G Hall, the Reverend Wade, the Reverend J J Robinson; and S F Gosling, Honorary Surgeon.

The official proceedings commenced with the review of the Corps by Adjutant Buchanan, who put the men through several drill evolutions, resulting in high praise by all present. They were then drawn up for the event of the day. Major Williamson, addressing Captain Williamson, said – "it was with much pleasure that he performed the duty which fell to him this day of presenting in the name of the donor, a silver bugle for use by the Kidsgrove Rifle Volunteers". In his speech he mentioned his involvement with the Volunteers of the 1798 era, remarking that he had been Commander of the Longport corps. Many formalities followed including the photographing of the whole corps by Mr Ludovico Hart of Burslem (who by 1877 had settled in Australia, and became a renowned photographer in Melbourne). At the invitation of Major Williamson, the men then marched to the Bleeding Wolf, Hall Green, for a sumptuous dinner also provided by the Major. As was customary, the evening ended with a series of pleasurable "Good Health's" and "Cheers".

It is evident from multiple newspaper articles that Adjutant Captain James M Buchanan was quite a popular man within the entire 1st Battalion. Before accepting the office of instructor of the Volunteers he had held a similar appointment in the regular army. It was he, that on 1 September 1860 (SSCGA) called together all the non-commissioned officers of the Battalion at the Railway Inn, Stoke, to deliver a lecture on the special duties of non-commissioned officers, imparting "much practical advice". Forty non-commissioned officers were present at that event.

In 1863 the 1st Battalion was under the command of Lieutenant-Colonel Buller, who was assisted by Major Hollins, Major Brown-Westhead, and Captain Buchanan, the Adjutant and Mr. Folker, the battalion surgeon. It was also reported on 2 October 1863 (SSCGA) that the annual meeting of the battalion for rifle was held on the previous Monday and Tuesday at the Newcastle butts (I hasten to add that a "butt" is a protective barrier, often a mud bank, that prevents stray ammunition flying too far past the target area, and has its origins in the bow and arrow days).

The muster of competitors at this Newcastle event was strong, and the results of the contests showed great improvement in their use of the rifle compared with the previous year. All the companies in the battalion were represented. Captain Buchanan was the chief superintendent at the shooting. In December 1860 (SSCGA), about 70 non-commissioned officers gave a dinner for their Adjutant Captain Buchanan at Hanley Town Hall. Regrettably, the death of this long serving Adjutant was reported on 8 December 1877 (SSCGA). He was buried at the Hanley Cemetery, with full military honours. The volunteers of Hanley, Newcastle (the two companies of each town), Stoke, Burslem, Tunstall, Longton, Stone, Leek, and Kidsgrove were present at his funeral.

On 28 December 1861 (SSCGA) it was reported that the Kidsgrove corps had organised a Ball, postponed until January 1862 due to "lamented death of the Prince Consort" (SSCGA). By 1862 there was a change in the command of F-Company with Captain W. S. Williamson stepping down in favour of Captain Edward Williamson, with Lieutenants Hugh W Williamson and Thomas Brogham, and Ensign Robert Clive. The Reverend J. J. Robinson and Hon. Surgeon Samuel Gosling remained in office. Incidentally, although described as Captain Edward Williamson he was more likely Robert Edward Williamson, the son of Robert Williamson Jr., and the widow Ann

Tinnersley. He would later take up the residency of Clough Hall, his mother's estate. Captain Edward Williamson was to remain in command of the Kidsgrove and District Company until 1876, a total of 14 years in Command, plus 2 years as a Leiutenant in the Corps.

The above changes in command reflected a quite general trend in the whole RVC Movement, which almost universally witnessed numerous Officer turnovers by 1863, though some individuals remained in a quasi-volunteer Officer Class (Ref 2). At this point we must confess to not being able to locate any detailed information on the rank-and-file members of the Kidsgrove and District Corps. The great majority of Volunteers in the urban areas were drawn from clerks, tradesmen, shopkeepers, publicans, and increasingly over time craftsmen and artisans, in some cases encouraged by their employers. Volunteers from non-urban areas like Mow Cop included farmers and farm labourers.

At the national level it was estimated that the level of income for an average Volunteer was about 18 – 20 shillings per week (Ref 2, Loc 1908). I suspect that this figure may not have been as high in the Pottery towns. In any event, the turnover in Officers was amplified in the turnover of rank-and-file members. By 1877 some 632,911 volunteers had passed through the ranks, exclusive of the 185,501 Volunteers enrolled at that time (Ref 2, Loc 1938).

Another notable example of the activities of the Kidsgrove RVC reported in the newspaper is that the Company attended, in force, the Annual Review of the Staffordshire Volunteers reported on 3 August 1867 (SSCGA). Five Battalions were represented at the Stafford Common with over 1,000 men reported present. Kidsgrove and District's representation was Captain Williamson, Lieutenant Brindley, Ensign Clive, Assistant-Surgeon Gosling, 5 sergeants, 1 bugler, 15 band and 60 rank and file – making a total of 84 men. As an aside, an Amateur Dramatic performance by members of the Stoke RVC was reported (SSCGA, 19 October 1867), yet another typical fund-raising public appearance for some members of an RVC.

3A. Some examples of Rifle Shooting Competitions

Among many of the positive outcomes arising from the Volunteer Rifle Movement was the massive growth in the numbers of individuals who could shoot with reasonable accuracy. The growth in Rifle shooters resulted in the formation of a plethora of Rifle clubs, both within the Volunteer Movement and outside it, for example, the Staffordshire Rifle Association. This brought with it numerous opportunities for inter-Company, inter-Corps and inter-Battalion Shooting competitions, held both locally and nationally; and participation in "Sham Fights", as they were called; and out of this organisation emerged awards of small cash or in-kind prizes at numerous "Prize Nights" held in RVC localities all over Britain. Such Award nights often took place in local pubs. The National Rifle Shoots of the RVC were frequently arranged to take place at Wimbledon, attracting Royal patronage.

Examples of the Rifle Shoots of the 1st Battalion were reported in 24 October 1863 (SSCGA), at the Newcastle butts over two days, with get-together luncheons served at the Albion Inn. A more typical internal Shoot (SSGCA, 2 March 1867) was a match reported between two sections of the Kidsgrove Corps - Mow Cop v. Kidsgrove, which Mow Cop won by one point. This report is one of so few that actually relate to the location of the butts, in this specifically at Harecastle Wood. The "after" event was held at what became the local haunt for the Kidsgrove F-company, the Harecastle Hotel, with dinner provided by Mr Poyser the landlord. Ensign Brindley, Sergeant-major Tomkys and other officers were present. It was noted that Captain Williamson was not well on this occasion. Another occasion at the Harecastle Hotel was the Company's Annual Tea in 11 December 1869 (SSCGA) where 150 attended including officers – the largest single gathering in Kidsgrove's RVC history. After tea the Band struck up and dancing went on to a "late hour".

It is also evident that the road between Kidsgrove and Harecastle Hill became well-trodden (marched) by the Kidsgrove corps and others. This road leads through Acres Nook and by an Inn there called the Rifleman. In his historical description of the Rifleman Inn the Kidsgrove historian Mr Philip R Leese has noted that this Inn became a major venue for the "after-shoot" such that it started to be called "The Volunteers". Lease also indicated that the Inn added stables to the premises, perhaps to accommodate the Officer's horses. Regardless, the name "The Volunteers" persisted for a period and is reported in the Staffordshire Sentinel (SS, 29 July 1879) as the site for the Kidsgrove Rose and Pink Show that year. Another contributor to this part of the story is Michael Moseley in one of his postings on Kidsgrove's Facebook page. He shows the advertising poster for the Auction of the Rifle Volunteers Inn on 18 January 1876. I feel obliged to comment about the Rifleman sign outside the Inn today (photograph), which appears to depict a rifleman dressed in the garb of the late 18th century, as opposed to the 19th.

As examples of Prize award nights for the Kidsgrove and District RVC, one event reported on 8 January 1870 (SSGCA) notes that on the previous Friday the Company met at their butts to compete for a number of modest prizes gratuitously supplied by friends of the movement. After the shooting the company retired to the Volunteer Inn at Acres Nook for an excellent dinner provided by Mr Cumberbach. The prizes were delivered by Mr Laurence Dale of the Roebuck Inn. First prizes went to Private Cumberbach and Sergeant Shortland. To my knowledge this is one of a few Press reports of the Rifleman Inn being referred to as the "Volunteer Inn".

Having broached the subject of "butts", I became curious about other butts mentioned in this work. I discovered an "Old Victorian Ordnance Survey Map, dated 1888-1913" (Ref 4) which provides some useful insights. By taking small sections of the Map, Harecastle Hill (just north of Ravenscliff) and Target Hill (just South of Bathpool) are easily identified as marked on the Map. Both sites are handily placed on the Acres Nook Road and provide a reasonable march from Kidsgrove or Goldenhill. On Target Hill a couple of measured target distances are shown, but curiously Target Hill is marked as Rifle Range "disused"? Maybe it's the age of the Map.

This raised my curiosity about other 1st Battalion Rifle butts. The most obvious one is at Bradwell Wood used by the Tunstall Company and shown on Map 2. The full set of distance measurements to target area are clearly marked. We also find a fully defined rifle range near Carmountside (Map 3) that may have suited the Burslem Company. On the other hand, Hanley was reported to be using new butts at Birches Head, but we draw a blank in this respect on the Survey (Map 4). However, another well-known target range was the one used by the two Newcastle Companies in the woods up towards Keele Hall from Silverdale and now buried under the houses of the Westlands estate. I was surprised to find an on-line essay by David Haden entitled: "On J.R.R. Tolkien at the Butts, near Newcastle-under-Lyme, in September 1915" (Ref 5). The renowned scholar Tolkien and his platoon were encamped close to these Newcastle butts and did their 3-to-4-day pre-WWI range training there. Even today, the view from the Butts (about 15 miles) is of the lower hills of the Staffordshire Moorlands, notably Mow Cop. It certainly makes one wonder if Tolkien's imaginative flair may have featured Mow Cop in some later poems or essays?

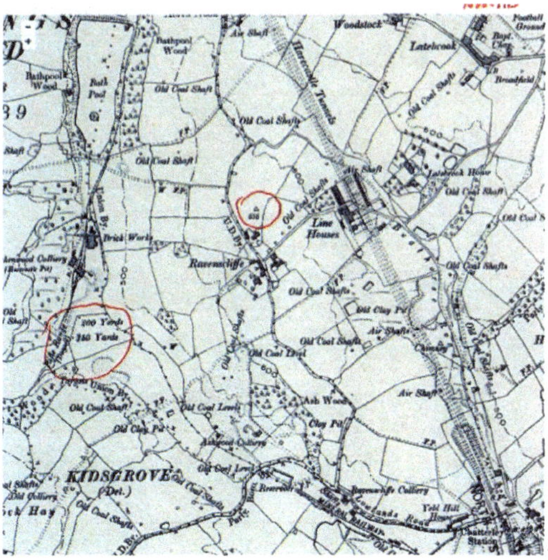

1

Chemical Works
Tank
Old Quarry
Bradwell Hall
Church View

2

L T O N
1397·381
Abbey
Meth. Chap.
White House
Old Shaft

3

4

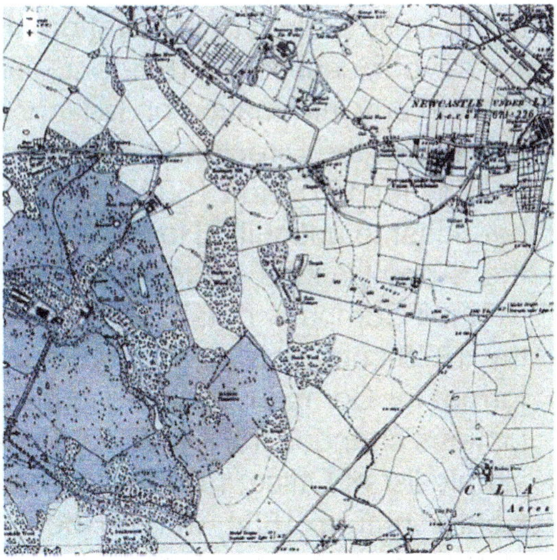

5

Old Victorian Ordnance Survey Map, dated 1888-1913
(Ref 4) Rifle Butt Sites

The "changing of the Officer Guard" for the Kidsgrove and District Company occurred in 1876 with Captain Robert Clive taking command, assisted by Lieutenant H. W. Williamson and Sub-Lieutenants Frank Hilditch and R. H. Gibbons. In the earlier section dealing with events in Astbury we cited a newspaper article (SSCGA, 11 September 1880) covering a divine service held in honour of Lieutenant Hugh W Williamson on his retirement from the Company, Frank Hilditch also resigned that year.

Sadly, for our purposes, the earlier editions of the Kelly's Directory of Staffordshire that we have accessed for 1872 and 1880 do not contain any information on Military activities, and two of the Directories in the series, notably for 1876 and 1884, seem to have been lost to antiquity. However, the format of the "Army Listings" up to 1884 are presented with full Company details as shown below for the 1878 entry. It is also unfortunate that the format for Company Officers in the Army Listings changes from 1884 to a relatively simple list of Officer ranks within the entire Regiment from 1884, with no identification of allocations to specific Corps.

However, the Regimental officer listings provided in the Army Listings we can track some old and new names of Officers. As examples of significance to this analysis Captain R Clive is listed as an Officer from 1876 – 1900. During his tenure as Captain of the Kidsgrove and District Corps the 1881 and 1882 Monthly Army Lists contain the promised heading “Staffordshire Rangers”, Stoke-on-Trent for the Regiment, but from thereon the Regiment is referred to as 1st Volunteer Battalion (late 2nd Staffordshire) Stoke-on-Trent. The Kelly’s Directory for 1888 confirms that Clive remains with the Kidsgrove Company, though promoted to Major. This same Robert Clive is made Hon-Colonel in 1892 and is second in command of the Regiment, and by 1900 is its’ Commandant. At this point Clive was awarded a distinguished Voluntary Decoration (VD), which was first awarded as an honour for Volunteers in 1892. A long-service medal for rank-and-file was introduced in 1893.

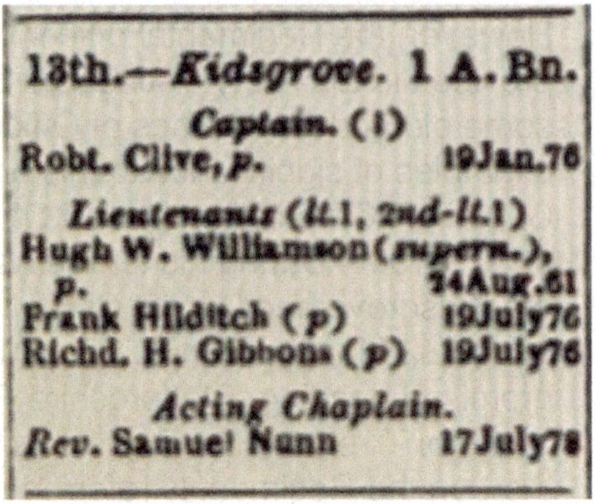

13th.—*Kidsgrove.* 1 A. Bn.

Captain. (1)

Robt. Clive, *p.* 19Jan.76

Lieutenants (*lt.*1, 2nd-*lt.*1)

Hugh W. Williamson (*superm.*), *p.* 24Aug.61

Frank Hilditch (*p*) 19July76

Richd. H. Gibbons (*p*) 19July76

Acting Chaplain.

Rev. Samuel Nunn 17July78

Army Listing 1878

Another person of significance to our analysis is W. H. Grindley, whose name appears first in 1888 as a Lieutenant in the Regiment. As of 1889 Captain W. H. Grindley is described as "Instructor of Musketry" for the entire Regiment, a position he retained in through to 1897. Most important is that the Kelly's Directory of 1896 shows that the 13th Corps F-Company has been relocated to Goldenhill with Captain W. H. Grindley in command; there is no reference to Kidsgrove and District. We can similarly track the fortunes of several other Volunteer Officers of significance to Company-F using the Monthly Military Lists including J. H. Clive, W. Tunnicliffe, C.R. Telputt and W. G. Laybourn. The Military Lists, which were originally stored by the Law Society of Great Britain, but were given over to Stanford University in the USA to be completely digitised – ideal for researchers based in different parts of the world.

Returning to the chronological story of Company-F events. A Kidsgrove RVC Prize award night was reported on (SSCGA, 24 December 1881). It was held at the Harecastle Hotel, the ceremony being "performed" by Mr T. E. Story of Kidsgrove, assisted by Captain R. Clive and a Lieutenant Mellor, who does not appear to be listed in the Kidsgrove corps. A substantial tea was served by Host Busbridge. Speeches extolled the view of "being prepared" for emergencies. A long list of prizes of 10 shillings each were given for attendance at drills, some for the lesser amount of three-quarters of that sum (7/6), but the greatest rewards went to a long list for shooting. I could not help but observe the names of an Alfred Caton and the Heaths, Dales, Durbers and Twaits, especially in thinking about some possible Goldenhill connections. I must mention Sergeant Brassington who through old age and lengthy service was retiring to honorary membership. Captain Clive ended by noting a meeting he had with the Inspecting General of the district who had remarked that F-Company was equal, if not superior, to any other in the Regiment.

As a general observation I note that even in the late 1870's the threat from France had not gone away, and I could not help myself when I found the below cartoon on the Channel Tunnel, which was commenced at the French side at that time.

THE PROPOSED TUNNEL.

LA FRANCE: "You're not really afraid that I will pop out of this hole when you're not looking, are you, John?" JOHN BULL: "Why, no, my dear; we're going to be much too good friends for anything of that kind to happen: and besides——" (He pauses and looks at ships.) LA FRANCE: "Precisely, John!"

3B. Brass Bands

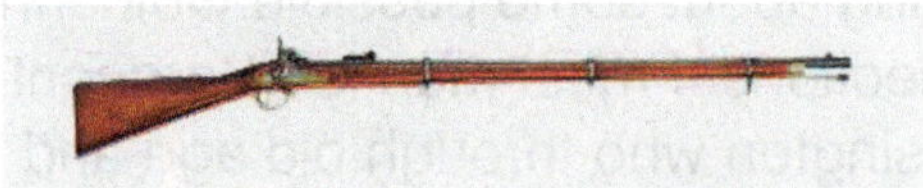

Having a Group of Rifle Volunteers on hand within the community provided many an opportunity for "Company" involvement in numerous local events, and to raise funds. This was even better if the Company had a Marching Band. In the case of Kidsgrove an SSCGA report of 3 April 1869 notes the Band in their new and handsome uniforms, offered a Grand Concert in the Assembly Rooms at Kidsgrove with Mr T Spencer as host. Regrettably, with the exception of officers, it was reported that there were few members of the corps present.

In contrast, at the Wakes Festival (SSCGA, 24 July 1869) members of the United Order of Oddfellows, respectably dressed, assembled at the Nelson Inn and formed a procession headed by the Kidsgrove Volunteer Band. They paved their way to Tunstall and "other places" (reporter's words), before returning to the Nelson where the landlord Mr Hancock provided them with an excellent dinner. Similarly, the Nottingham Order of Oddfellows held their anniversary at the New Inn headed by the Tunstall RVC Band. In this report I discovered a completely new game to me – called "Cockup" or "Coppuck" where men apparently wore the ugliest of masks and effectively went on a pub crawl.

It should be obvious by now that quite early in the development of the Volunteers

Movement, Brass Bands and Brass Band contests emerged in the form of district and national events. For example, on 21 August 1859 (SSCGA), the 1st Battalion of the Staffordshire Rifle Volunteers organised a Contest at the Park at Port Hill. There were charges for admission, and a pre-sale of tickets at the Warwick Savage Bookseller in Burslem. This period represented the beginning of the Brass Band era, which was really a response to industrialisation.

Technological advances in musical instrument design, and in mass production and distribution had occurred. It became relatively inexpensive to form a Band of up to 26 to 28 members, although a smaller number could suffice (in the early days of the Kidsgrove and District example the Band comprised 15 players.

As a youngster growing up in Goldenhill, close to a hundred years later I remember the Saturday afternoon BBC radio being devoted to Brass Band Competitions – the beginnings of such events were clearly much earlier in history. But then, who could forget the renderings of the Black Dyke Mills Band, the Foden's (truck manufacturer), Fairey (aircraft manufacturer), and Leyland (vehicle manufacturer) Bands, or the many coal mining district bands like the Grimethorpe Colliery Band and the Carlton Main Frickley Colliery Band, all born out of the 19th Century Brass Band movement, going strong in the 20th Century, though regrettably I am told not as popular in the 21st Century.

It was not uncommon for the RVC Bands to be involved in providing performances at the local schools, for example, in May 1870, the Kidsgrove Band gave a successful concert at a Goldenhill Schoolroom, aided by the Tunstall Glee Union. During the Dahlia Show (SSCGA, 16 Oct 1869) of the Pitshill and Goldenhill Floral Societies at the Crown Inn, it was reported that the Kidsgrove Band gave their annual concert at the Kidsgrove Assembly Rooms. Similarly, the Ancient Order of Foresters were entertained by the Kidsgrove Band on 30 October 1869 (SSCGA). I must also add the festivities, including Band, for Mrs Kinnersley's Birthday party at Clough Hall (SSCGA, 23 July 1870), arranged by Captain Edward Williamson of the Kidsgrove Corps.

I could go on describing the numerous activities of the Kidsgrove Rifle Corps and others including sham fights, regimental drills (usually at Trentham), encampments, dinners, rifle shoots, parades, and more general get-togethers in one pub or another, however, none appear more important than the ceremony to honour Captain Edward Williamson on 1 July 1876 (SSCGA). The men of the 13th Company presented their retired commanding officer with a magnificent sword and belt with the inscription: Presented to Captain Edward Williamson by the present and late members of the 13th Rifle Volunteers Corps on his earlier resignation of the command on January 18, 1876, to show their high regard and mark their appreciation of his undying exertions for the welfare of the corps during a period of more than 16 years. The presentation took place at Ramsdell Hall. That retirement marked the beginning of the Robert Clive Era.

Having mentioned Sham Fights, I judged that one example reported on 22 May 1869 (SSCGA) may suffice: The "big" Sham Fight between the First Battalion Staffordshire Volunteers (1074 men divided into two battalions) and a smaller Cheshire Battalion (624 men, also divided into two battalions) held at Somerford Park, Congleton. The mimic battle had the Cheshire men attacking Congleton, while the Staffordshire men defended the town. All participants had to face a pitiless rain storm – they all became drenched to the skin, but it was the commanding Colonel of the Staffordshire Regiment that took pity on his poor soaked charges in ordering abandonment of the field. Unfortunately, it appears that he did not tell anyone else he was doing this – the Cheshire men certainly did not know and carried on regardless. Lieutenant-Colonel Sir Charles Shackerley (Cheshire) addressed the men at the end of event, humourlessly remarking that he supposed the Staffordshire men had been washed away. Needless to say, there were drinks on hand.

During the period 1872 -1881 a considerable number of significant Military reforms were introduced. This was all part of the increasing Government control used to compensate for the failure of the Volunteer Movement to be self-sustaining. The earlier Cardwell reforms had a devastating effect on the morale of the officers of the Volunteer Movement, resulting in nearly 2,000 resignations between 1872 - 74, that even after re-appointments resulted in a net loss of 400 officers (Ref 2, Loc 2395). The later Childers reforms focussed more on combining the Volunteer forces as an auxiliary to the Regular Army. In reality the Volunteer Battalions effectively became part of the Regular army and could thus be called upon to send reserves to serve as replacements in the regular army frequently based overseas, and often engaged in battle. The 1st Battalion was granted the right of using the name “Staffordshire Rangers” in 1881, although its formal title and commanding officers continued to be reported in the Kelly’s Directory as: Prince of Wales’s North Staffordshire Regiment, 1st Battalion. Commanded by Colonel Michael Daintry Hollings; Lieutenant-Colonel W H Dutton; Major B Moore; Captain G.L J. Goff, Adjutant; Captain Henry S Swann, quartermaster; Surgeon Major William Dawes; J P Edwards, honorary assistant surgeon; C H Phillips, acting surgeon; Rev J Watson and Rev S Nunn, acting Chaplains.

Four years later (Kelly’s Directory 1892) a large number of the names had changed, notably the Commandant was the Hon Colonel W H Dutton, with second in command the Hon Colonel R Clive; Majors are W W Dobson and H Palmer; Captain W H Grindley, instructor of musketry; Captain Goff remained as adjutant as did Henry S Swann as quartermaster and the Reverends Watson and Nunn as acting chaplains. A completely new set of medical officers (surgeons) were identified. Significantly there is no command staff for F-Company

Kidsgrove listed in the Kelly's Directory of that year.

Photograph of a member of the Burslem Rifle Volunteer Company-C, William Arthur Jones, between the period 1883 – 1890. Courtesy of the Trustees of the Staffordshire Regiment Museum

Chapter 4. A Military Presence - The Goldenhill Solution

At this point I hope the reader recalls the sudden absence of the F-company Kidsgrove & District from the 1st Battalion listings of 1892. I summarise this as the possible beginning of a "Gap" in the Table of command succession shown below; most certainly from 1888, since up to that date the Monthly Army lists continue to show Captain Robert Clive as commander. Clearly something had occurred in the Kidsgrove corps, and this was amplified following the promotion of Robert Clive to Major and his move to the Tunstall Corps in 1888. Unfortunately, there appear to be few clues in the local Press. Consequently, we have to place added reliance on the only practical evidence we have through the Military Army Lists and the Kelly's Directory of Staffordshire. Names that recur in the Military lists include Captain W H Grindley and Lieutenants J H Clive, G W Laybourn, and C R Telputt.

Succession of Command of the 13th Corps

Years	Command	Location
1860 – 1861	Captain W S Williamson	Kidsgrove
	Lieutenant Edward Williamson	
1862 – 1876	Captain Edward Williamson	Kidsgrove
	Lieutenant H W Williamson	
1876 – 1888	Captain Robert Clive*	Kidsgrove
	Lieutenant H W Williamson (to 1881)	
GAP		
1896 – 1900	Captain W H Grindley	Goldenhill
	no sub-command	
1900 – 1904	Captain G W Laybourn	Goldenhill
	Lieutenants H and L Clive (1904 only)	
1905 – 1908	Captain Harry Clive	Goldenhill

*Promoted to Major in 1888, moved to Tunstall Corps

As it happens, the "Gap" is embraced by a tragedy in world economic history, a major recession that ran from 1873 to about 1896, then known as the "Great Depression", but later modified to be called the "Long Depression", following the even more calamitous economic depression events of the 1930's.

The "Long Depression" began with the "financial panic of 1873" involving stock-market crashes, commencing in Vienna and London, with the collapse of the international market pumped up by a wave of speculative money aimed at the growth of railroads and real estate in the USA. Back in those days, railroads were a new invention, and US companies were borrowing money to get the cash they needed to build new lines. The slump that followed was caused by the build-up of debt, the arrival of new technologies, and an emerging USA economy that was flooding the world market with cheap goods (a bit like the China of today).

The initial depression lasted from 1873 – 1879, but then there followed a longer period of price deflation, though most businesses maintained their levels of production. For some businesses the level of borrowing increased, but profitability for industrialists and their financiers alike, declined (Ref 14). Price deflation was particularly significant in the raw material/mining and cotton manufacturing sectors over the period 1882 – 1885, and beyond in different parts of the world.

In Great Britain in industrial revolution districts like Kidsgrove and Goldenhill, the prices of coal and iron ore plummeted, and for many smaller operations profitability eventually collapsed. Britain's national wealth fell from an average of 12.8 percent between 1870 – 1874 to 9.7 percent between 1875 – 1896. In Australia, the impact was a little later, but saw Gross Domestic Product fall by 17 percent over the period 1892 and 1893 – the worst fall in Australia's economic history.

The "Long Depression" brought down the coal and iron masters of the Williamson family in North Staffordshire. The Williamson Brothers (Hugh Henshall and William Sheppard Williamson), owners of the Stone Trough Colliery Company at Scholar Green; the Brown Lees Colliery; and the Goldendale Iron Company, Goldenhill, filed for bankruptcy with liabilities exceeding £50,000 in November 1886 (Birmingham Daily Post, Thursday 11 November 1886), as also noted in the Mow Cop Interactive History site (Ref 11). Similarly, Edward Williamson (former long-term commander of the Kidsgrove and

District Rifle Volunteers), owner of Kinnersley and Company and its' Clough Hall Works filed for bankruptcy in 1887 with liabilities of £160,077 (Birmingham Daily Post, Wednesday 14 December 1887). These declarations of bankruptcy resulted in pit closures and several hundred miners in the KIdsgrove and Goldenhill district being made redundant.

The Staffordshire Sentinel of 10 October 1889 also reported that legal proceedings were issued against Hugh William Williamson to vacate his lease of Ramsdell Hall, the long-standing residence of the Williamson family. Ramsdell Hall was actually owned by the Chadwick-Lowndes family and had been leased to the Williamsons from the time of Robert Williamson Jnr. occupying the premises. It was Thomas Chadwick-Lowndes, wine and spirit merchant, and I believe a Solicitor, of Congleton, that was looking to repossess the family property, especially in the light of the insolvency of this part of the Williamson family. It appears that the lease of Ramsdell Hall may have continued in the hands of John Henshall Williamson (Ref 15), but we have been unable to ascertain the date when this commenced.

The Williamsons had been the stalwarts of the Kidsgrove and District Rifle Volunteer Company-F from its' commencement in 1859. It is thus tempting to speculate that one of the main issues for Company-F during the period of the Long Recession may have related in part to the inability of the Corps to find sponsorship funding; and yet in spite of the financial crises Company-F appears to have been sustained by its' Commander Captain Robert Clive through to at least 1888, Table 1 (page 17) from Talbot's work shows representative expenditures of the entire 1st Battalion Regiment significantly for the eighteen months from mid-1884 to the end of 1885.

Another source for the apparent demise of the Kidsgrove and District Company-F may reside in the difficulties associated with the continuous cycle of recruitment to the Corps over time. Philip Leese kindly informed me of a press clipping that he possesses from The Guardian Newspaper of Wednesday 1 April 1885 that covers "a Complimentary Dinner to Mr Proctor" (Mr Ephriam Proctor, then Chair of the Kidsgrove Local Board). The dinner was attended by a number of Rifle Volunteers including Mr Hook, Lance Corporal Douglas from the University of

Oxford Volunteers, and Sergeant-Instructor Isom (Kidsgrove).

Sergeant Isom spoke about the difficulties of recruiting new volunteers, especially from Kidsgrove. He noted that the young men of Kidsgrove seemed to be above joining the Corps, and that there seemed to be an increasing public dislike of “a man in a volunteer uniform”. In fact, he remarked that the public often made fun of such men. Sergeant Isom analysed the composition of the Corps over the 26 years of operations. He noted that 656 men were on the corps' books as having been “trained and drilled”, but only 127 of these were from Kidsgrove; and of the 14 volunteers recruited that year (1885), only one was from Kidsgrove. (As a general observation the author notes, however, that this was not solely a Kidsgrove Corps, but one representing the “Kidsgrove and District”). In any event, there was a local recruitment problem.

Whatever the ultimate source of the problem for the Kidsgrove and District Corps, something had snapped in the organisation, and the result was that the Kidsgrove and District F-Company became dysfunctional. As the issues of the price deflation episode slowed down, the opportunity for re-establishment of the F Company was found not in Kidsgrove, but in Goldenhill. This may have involved other members of the Williamson family, particularly John Henshall Williamson (father and son), whose major investments were in minerals processing for applications in the pottery industry through their Colour Works in Goldenhill.

In following the promising re-establishment path for Company-F, the initial problem was where in Goldenhill? A good deal of the village was demolished in the 1960s. We searched high and low for anyone who could perhaps throw light on where the Company-F 1st Battalion could possibly have been based in the village - a barracks, a drill hall, and/or an armoury? The latter two being a requirement for the operation of an RVC Company. An early view was that the area well-known in Goldenhill as "The Barracks" might have been the place, but a survey of the Census's for much of the critical period shows that the houses in this area were being used for the purpose for which they were designed – housing for working families, especially those working at the Williamson mines. My good friend Jon Goodwin in Stoke then stumbled across an interesting note in the Staffordshire Sentinel of 27 July 1903 that describes how the teachers and scholars of the Christ Church Sunday School numbering about 230 spent the afternoon in the "Drill Hall", which at that time was apparently used by the Sunday School.

I recalled that this was the site at the end of Rodgers Street (then Williamson Street) that was not only occupied by a single reconstructed Church building, but in my memory, also by a larger barn-like building that served the purpose of providing many of us with school meals in the late 1940's. For some reason I also recalled that the interior of the building had a small stage at one end. What were other people's memories of this big-barn of a place? In 1874 John Henshall Williamson (1814 – 1883) had bought and imported from Liverpool a corrugated-iron church building (the "Tin Church"), initially known as the Free Church (later Christ Church), to a site he owned at the top end of Rodger's Street. (To be honest I disliked going there when I was a youngster because I hated school meals). But, if the "ugly big barn" where school dinners were served, was a former Drill Hall, surely someone must have heard of its' original function?

After asking others in the Goldenhill Memories Group, initially to no avail, I thought I would ask my former-wife who lived on the "Levels", nearby the Tin Church in her early childhood and attended the Tin Church Sunday School. Without informing her of why I was asking the question I simply said: Do you remember ever hearing of a building called the "Drill Hall" in Goldenhill? She had no hesitation in saying - Yes! It was by the Tin Church. Unprovoked she went on to say that the building was used mainly for Wedding Receptions when she was a child, but there was a smaller building close by, which the children were always told to stay away from – she used the word "banned" from going near. Could this have been a former Armoury?

Subsequently David Salmon of the Goldenhill Memories Group confirmed that the Sunday School building at Christ Church was also split internally into two rooms. The larger room was entered initially up steps from Rodgers St. The senior Sunday school, pantomimes, bring and buys, dances, whist drives and all functions were held there in the late 1940's to 60's. There was a stage at the far end. There was also a small kitchen to the left. The smaller room is where the infants Sunday school was held at the rear of the building.

Additional newspaper searches finally revealed something quite exciting. A report in the Staffordshire Sentinel of 1 March 1889 describes a Volunteer Concert: A "smoking concert" held on the Thursday night in the "Gymnasium attached to the Armoury" at Goldenhill. (Jon Goodwin informs me that such "smoking concert" events were only open to men and persisted into the late 20th century). A second article (SS, 14 December 1889) called "Kidsgrove Concert" describes an event in aid of the prize fund for the F-Company, 1st North Staffordshire Battalion that was held at the Drill Hall, Goldenhill on Thursday night before a crowded and enthusiastic audience.

Let me remove the suspense, Armoury is defined clearly, Gymnasium = Drill Hall more than likely. A quick look at an old photo of the Tin Church shows there were three buildings on the J H Williamson site even as late as 1964. Behind the church were two buildings: a small building probably the "Armoury", and the long building the "Drill Hall". You can see the Drill Hall in the current Header photograph of the Goldenhill Memories Group Facebook page, featuring the old PMT garage. Of course, most of these buildings with the exception of the church no longer exist, following the latter's rebuilding and the mass demolition of the old village.

Alan Goodwin of the Goldenhill Memories group also supplied a site Map for the Church grounds in about 1888, which clearly shows the entrance to the "big barn" from Rodger's Street.

To round off, a third Staffordshire Sentinel article of 12 January 1889 describes a lecture on "Central Africa", given on a Thursday evening at the "Drill Hall", Goldenhill by Lieutenant Telputt of the Kidsgrove Volunteer Corps. Captain Grindley was present, in recovery following his recent convalescence. The reason this article caused me some amusement is because the Kidsgrove Corps was well established in Goldenhill by that time; and secondly the article is only a small note at the end of a very long one covering an horrendous murder trial described as "The Cross Hall Tragedy" (worth a read in its own right). The Cross Hall story is in fact a novel, the copyright of the New Zealand Herald (what a small world we really live in!).

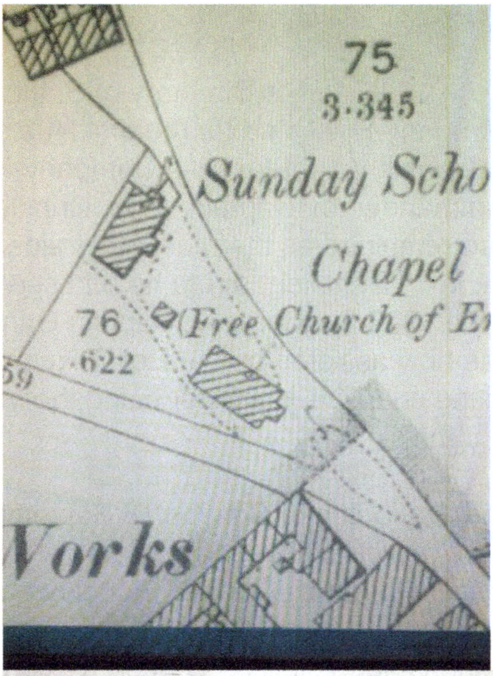

The Church and Sunday School at Christ Church,
Rodger's Street, Goldenhill

In the 1896 Kelly's Directory of Staffordshire, Company-F is listed for the first time as the Goldenhill Volunteer Rifle Company-F, commanded by Captain W. H. Grindley, who also remained as the musketry trainer for the entire 1st Battalion, a role he had served since 1892. Incidentally, Captain Grindley was ably assisted by the Rev T Nunn as acting chaplain, the latter being the Vicar of St John's.

The Staffordshire Sentinel, 8 September 1897 reports a meeting held at the Red Lion Hotel, chaired by Captain Grindley and convened to present Prizes to the members of the Goldenhill RVC Company-F primarily for drill attendance. As a synopsis, Colour-Sergeant W R Green, Private T Kennerley, and Lance-Corporal W Edwards were the top three in drill attendance, but another 18 received awards (see the following "clipping"), The presentations were followed by the usual salutations. Sergeant Shirley, who also accompanied other vocalists, rendered a number of songs. You may even find an ancestor in the list of winners – if you do please let me know.

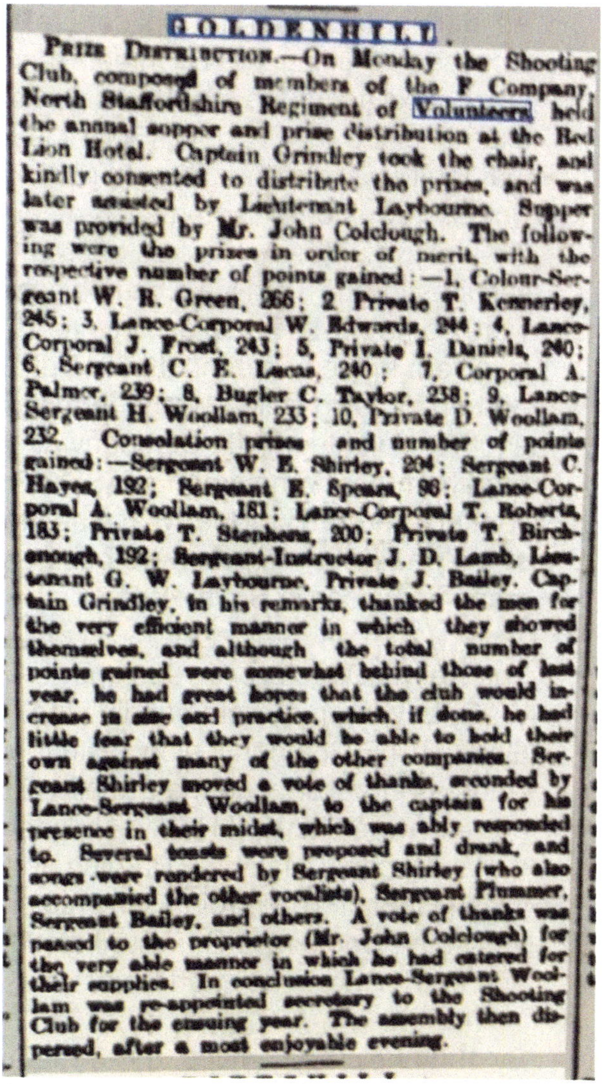

GOLDENHILL.

PRIZE DISTRIBUTION.—On Monday the Shooting Club, composed of members of the F Company, North Staffordshire Regiment of Volunteers, held the annual supper and prize distribution at the Red Lion Hotel. Captain Grindley took the chair, and kindly consented to distribute the prizes, and was later assisted by Lieutenant Laybourne. Supper was provided by Mr. John Colclough. The following were the prizes in order of merit, with the respective number of points gained:—1, Colour-Sergeant W. R. Green, 266; 2, Private T. Kennerley, 245; 3, Lance-Corporal W. Edwards, 244; 4, Lance-Corporal J. Frost, 243; 5, Private I. Daniels, 240; 6, Sergeant C. E. Lucas, 240; 7, Corporal A. Palmer, 239; 8, Bugler C. Taylor, 238; 9, Lance-Sergeant H. Woollam, 233; 10, Private D. Woollam, 232. Consolation prizes and number of points gained:—Sergeant W. E. Shirley, 204; Sergeant C. Hayes, 192; Sergeant E. Spears, 96; Lance-Corporal A. Woollam, 181; Lance-Corporal T. Roberts, 183; Private T. Stenhens, 200; Private T. Birchenough, 192; Sergeant-Instructor J. D. Lamb, Lieutenant G. W. Laybourne, Private J. Bailey. Captain Grindley, in his remarks, thanked the men for the very efficient manner in which they showed themselves, and although the total number of points gained were somewhat behind those of last year, he had great hopes that the club would increase in size and practice, which, if done, he had little fear that they would be able to hold their own against many of the other companies. Sergeant Shirley moved a vote of thanks, seconded by Lance-Sergeant Woollam, to the captain for his presence in their midst, which was ably responded to. Several toasts were proposed and drank, and songs were rendered by Sergeant Shirley (who also accompanied the other vocalists), Sergeant Plummer, Sergeant Bailey, and others. A vote of thanks was passed to the proprietor (Mr. John Colclough) for the very able manner in which he had catered for their supplies. In conclusion Lance-Sergeant Woollam was re-appointed secretary to the Shooting Club for the ensuing year. The assembly then dispersed, after a most enjoyable evening.

Press clipping from The Staffordshire Sentinel 20 January, 1897

As mentioned earlier the Childers reforms focussed on using the Volunteer Regiments as an auxiliary to the regular army. For example, (SS, 20 January 1900) it is reported that the Mayors of the different participating Pottery towns organised a send-off dinner to the local volunteers who had offered their services for the Transvaal campaign. There were some 60 1st Battalion volunteers involved. In the same month the Duke of Sutherland also called for the formation of another Volunteer Rifle Company at Trentham "for home defence" (SS, 13 January 1900). The Boer War was then in full swing and British troops were being sent to many provinces in South Africa.

The Kelly's Directory of 1900 shows Captain W. G. Laybourn as commanding the Goldenhill F-Company, along with the Rev Dunn as chaplain; and the Battalion staff officers were Hon-Col R Clive, Leiutenant-Col W. W. Dobson, Majors J. Harding and W. E. Clark, Captain B. Wardle as instructor of musketry, Captain H. G. Powell as adjutant, and G. H. Swann as quartermaster.

By 1904 help had arrived at F-Company, Goldenhill with Captain Laybourn having two Lieutenants Harry Clive and Laurence Clive added to his squad. There are clearly some inconsistencies here since on 27 April 1904 the Staffordshire Sentinel reported that Captain H Clive addressed the men as their new Captain since Captain Laybourn had retired. He praised Captain Laybourn stating that everything he did was for the improvement and comfort of each member of the Company.

Some of the meetings of the Volunteer Rifle Companies then commenced to revolve around something referred to as "Annual Social Club Suppers". On 1 October 1903 the Staffordshire Sentinel reported that F-Company, Goldenhill held its annual club supper at the Nelson Hotel. Lieutenant H Clive presided, supported by Lieutenant L Clive and Messer's Pigot, Riley, Hassam, Wood and Davidson were present. The report notes that the only toasts were to "The King" and "the success of the social club". There followed singing. At around this time I also came across a report (SS, 27 November 1903) which describes a Morris Tube Competition for which there is a League Table with each Company listed. The Morris tube is a removable liner inserted in the barrel of the rifle that enables the use of .22 ammunition as an option to the standard .303. These were introduced to encourage the use at small indoor rifle ranges, for example, in a suitably equipped Drill Hall.

The annual Volunteer supper (SS, 27 April 1904) mentioned above had involved a parade through the streets of the village by 100 Goldenhill Volunteers. But, perhaps of greater importance was that Captain Clive brought up the subject of an "important body", in fact a Royal Commission, drawn up to consider the future of the Militia and Volunteer forces. He stated that no one knew what the recommendations would be, but whatever else they were, the Volunteers must be part of the Army of Great Britain. There was much talk of efficiency and physical training. Clive believed that "before long the Government would insist that every Volunteer force should be judged alone by its power to contribute to a force of soldiers "fit to take the field in time of war". From the article by Philip Talbot (Ref 10) it is clear that the strength of the 1st Battalion was almost as large as it had ever been (see Table 7) at 1,357 men. (The numbers of members in the corps designated in the Table for Kidsgrove (101) in 1900 are clearly for the Company-F, based in Goldenhill).

Captain Clive noted that the Company only had 11 more drills before camp, and he urged everyone present to do his best to attend all of these drills. He wanted them to make an extra special effort, as was also the wish of Lieutenant L Clive to bring the Company to a high standard of “efficiency”. It was at this point that Lieutenant L Clive intimated that Colonel Dobson had offered him the command of the Tunstall Company.

TABLE 7

1st Volunteer Battalion North Staffordshire Regiment: Composition 1900

Company	*Location*	*Strength*
A	Longton	101
B	Hanley	99
C	Burslem	88
D	Tunstall	93
E	Stoke	113
F	Kidsgrove	101
G	Newcastle	89
H	Newcastle	81
I	Leek	85
K	Hanley	91
L	Stone	96
M	Trentham	95
N	Eccleshall	70
O	Hanley	81
P	Biddulph	74
	Staff	8
	TOTAL	1357

In a Press report (SS, 4 May 1904) we learn that both Captain Meakin and Captain Laybourn had resigned as Captains of

Tunstall and Goldenhill, respectively, and that the Clive cousins had been appointed to replace them. The words "increased efficiency" kept appearing in reports on the thoughts of the cousins. I believe the "writing was on the wall" for the Volunteers at this point, but let's keep up the suspense.

By 24 December 1904 (SS), the Volunteer Prize distribution for the 1st Battalion F-Company as reported as being held in their Drill Hall, Goldenhill on the previous Thursday evening with supper provided by Captain Harry Clive. I note "the by-no-means beautiful building had been gaily and effectively decorated by extensive use of flags and bunting". The occasion gained interest by the presence of the Regiment's Colonel B C Clive. Captain Harry Clive called upon his father the Colonel to present the prizes, noting that the number of men who went to camp this year was decidedly better than the previous year and that shooting was better. 48 prizes were presented for attendance and 23 shooting prizes to: Sergeants Green, Spear, Brough, and Farrall, Lance•Corporal Woollam, Private Johnson, Corporal Harrison, Sergeant Painter, Private J. Ferrell, and Sergeant H. Woollam. The first shooting prize in the recruit elites was won by P. Bailey. (You may have a relative amongst these names). Mr. H. H. Williamson (the 2nd) proposed a toast to the 'Prosperity of the Company." Mr. W J. Hassam gave the toast to the "Health of the Chairman".

In the Staffordshire Sentinel of 24 December 1905 is a report of the annual supper, prize distribution, and concert held in the Drill Hall, Goldenhill. This was the sixth year that Captain Harry Clive had prepared a report on the year's work of his Company which then numbered 84 men with 82 going to camp. He commented that few men had any difficulty with the medical at the camp, although five were certified as unfit. Captain Clive then expressed great pride that 41 of his men had attended every parade. Long service medals were presented to Colour-Sergeant Lucas, Sergeant Harry Woollam, and Private C, Morris. Colour-Sergeant Lucas also received a gold chain and illuminated certificate for his 20 years of service. Musical entertainment followed the presentations.

A similar annual dinner/presentation for the F-Company occurred on 4 October 1906 (SS) held at the Nelson Hotel, Goldenhill, where "Cyclist-Sergeant Painter" provided dinner. Following dinner, a smoking concert was held, including songs by Captain Clive,

Colour-Sergeant Oargrave (flute), Sergeant Bough, Corporal Bullock, Private Madden, Messers Anderton, Ankers and Worrall (violin) with Mr Anderton the accompanist.

Later in the year (SS, 22 December 1906) there is a report of a strong muster of F-Company at the Drill Hall, Goldenhill for the annual prize distribution. A hot-pot supper was served by Cyclist-Sergeant Painter, catering once again. Captain Clive reported that the Company had 79 trained men and another 30 recruits. At camp not a single man had fallen out of their 30-mile march, He particularly thanked the non-commissioned officers for their contribution to this success.

Mrs. Clive then presented the prizes to: Sergeant Bough, Sergeant A. Farrell, Lance-Sergeant Horrocks, Sergeant Green, Cyclist-Sergeant Slater, Sergeant A. Woollam, Corporal Woollam, Private Lamb, Colour-Sergeant Woollam, Corporal Horrocks, Private Nixon, Private Williams, Private Cotton, Lance-Corporal Roberts, Private Lees, Lance-Sergeant Smith, Privates Murray, Frost, and Bailey, Corporal Bullock, Privates Farrell and Jones, and Corporal Farrell. The recruits' prize was won by Private Ball. A musical evening followed.

1907 - enables us to explore a different series of events, but life goes on for F-Company, Goldenhill. A report (SS, 1 March 1907) contains a short extract concerning an F-Company Morris Tube Club annual dinner on Tuesday 26 February at the Lord Nelson, Goldenhill with 48 members and friends present for a meal provided by Cyclist-Sergeant Painter, followed by a smoking concert, the artists being Messer's Grover, Mulroy, Andrews, Ankers, Sergeant Brough, Omen Painter and Private Madden with humorous sketches by Mr Hermon.

We have mentioned Cyclist Sergeant Painter several times. He is representative of the new recruits to the Volunteer movement at that time. The Regular Army was in fact highly supportive of the Volunteers trying out new ways of doing things so here is a story from the Staffordshire Sentinel on the Sergeant called: THREE MILE CYCLE RACE:

A most exciting three-mile bicycle race was witnessed by over 1,000 people on Monday evening on the Clough Hall track. The race was between E. Hughes, of Kidsgrove, and T. Painter, of Goldenhill, for a gold medal, and prize value £l0. Hughes, who held an unbeaten record, over various distances, in several previous competitions on road and track, was conceding Painter a start of 300 yards. A considerable delay was caused through the track being six yards over the quarter mile when measured. Hughes and his supporters would not give way to this being taken into consideration, and there was a doubt if the race would go on, until Painter, in a sportsmanlike manner, decided matters by foregoing and risking the difference it would make to his winning chance by lessening his start from 300 to 228 yards; in order not to disappoint the assembly. Painter had been the universal favourite until this occurred, and on the top of this, had the bad luck to lose the toss in the choice of the starting point, a great feature on this rough, neglected track. A good start was affected at 6:10 p.m., Hughes gaining tremendous ground in the first two laps. Painter, who is a fine athlete, rode in good style, but was going very steadily up to this point. Then suddenly he seemed transformed, and with terrific speed he regained the lead in one lap, and amidst great excitement never slackened until the seventh lap, when he was rewarded for his efforts by being close up behind Hughes, almost a lap in front, and now riding hands-up for the rest of the distance, won a most popular victory, for the public who shouted themselves hoarse in showing their appreciation. Painter who was trained by George Cooke, late of St. Helens, and then trainer to Port Vale F.C., is a cyclist-sergeant in the 1st Volunteer Battalion N.S. Regiment Cyclist Corps, and also the popular host (Landlord) of the Nelson Hotel, Goldenhill.

If you wish to read more about the "cycling corps" I recommend the paper describing the 26th Middlesex Cyclist Volunteers (Ref 12). The Cycling Corps were abandoned in 1920 to become the Signallers.

The largest event on the 13th Corps reported in 1907 was unquestionably on 25 June 1907 (SS). This article describes a divine service at the Parish Church, Goldenhill for members of the Tunstall Council. The Company assembled at the Volunteer's Drill Hall, Goldenhill and formed into a procession. The bugle band of the PFT Division of the St. John Ambulance Brigade went first. The Volunteers including members of the cycling section, drawn from the F-Company (Goldenhill) and the D-Company (Tunstall) of the 1st Volunteer Battalion North Staffordshire Regiment in the command of Captain Harry Clive followed, and more. The church was filled to capacity.

I failed to mention that the Brass Band associated with the earlier Kidsgrove group had not transferred to Goldenhill as far as I can tell, probably on the grounds of cost. We are aware, however, that in 1879 an enlarged "Kidsgrove Brass Band", had performed at the Albion Inn and was apparently going strong in its independence of military involvement.

The Staffordshire Sentinel on 28 December 1907, reports what starts to look like the final Prize distribution and "smoker" for the F-company, Captain H Clive commanding, supported by Captain Furness, RAMC. Captain Furness pointed out to the audience that the Boer and Japanese campaigns had shown that more Artillery and Cavalry and less Infantry would be needed in future wars. He was also in favour of the Secretary for War being a long-term role, not changed by every election, and he personally was in favour of conscription. During the evening Messer's Ankers, Brinder, Maddon, and Mulroy entertained the gathering.

Chapter 5. The Final Demise of the Rifle Volunteer Movement

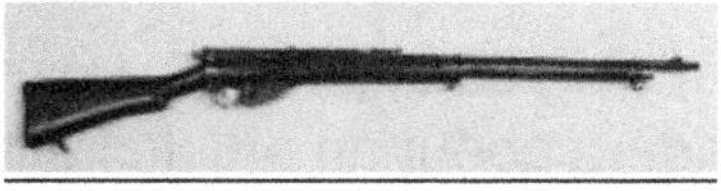

The Kelly's Directory of 1908 introduces in many ways a repeated problem (we do not know precisely the date of publication). However, the Military listings show a Staffordshire Volunteer Infantry Brigade including the 1st Battalion Prince of Wales's North Staffordshire Regiment, but as if history is repeating itself, the F-Company has once again disappeared.
A clue is found in the listing for the D-Company, Tunstall where Captain Harry Clive (formerly of Goldenhill) is listed as in command. This article (SS, 14 February 1908) also notes that the Volunteers in North Staffordshire are at present passing through a "big crisis".

A scheme developed in the War Office provided for only 4 Battalions of infantry in the county of Stafford from the five that existed. Several meetings of officers to decide which Battalion would be disbanded took place and among the other schemes proposed was a proposal to convert the 1st

Norths into an artillery corps. This was averted at that time, but which companies would be disbanded remained to be seen.

Stepping back to 1905, I cannot avoid telling readers about an article in the Sheffield Telegraph reported in the Staffordshire Sentinel of 19 January 1905 which says: There are strong reasons for the belief that radical changes will shortly be made in the rifle arm of the volunteer's service; and there is considerable unease in at least two Yorkshire Battalions, both of which were "under-strength".

We need to jump to the Kelly's Directory for 1912. Here we find the Royal Commission of 1904, and the Furness predictions of 1907 about army structure have been enacted. The Military lists for Staffordshire include the Royal Artillery 1st Staffordshire Battery, with Companies by town, notably Company-D, Captain A Clive, Tunstall; and Company-F, Lieutenant H A Ridgeway, Stone, with Longton, Hanley, Burslem, Stoke, Newcastle, and Trentham listed as the other companies in this Battery.

The 6th Battalion Prince of Wales's North Staffordshire Regiment is based in Burton-on-Trent, and the 5th Battalion South

Staffordshire Regiment in Walsall. In 1912 there remained one small Infantry group, the 5th Battalion Prince of Wales's Regiment North Staffordshire Regiment based in Hanley, but with no Company allocations. I remember as a teenager, threatened by Military conscription, the prejudice we all felt against joining an infantry Regiment.

In 1912 the J. H. Williamson's (1st and 2nd generation) Goldenhill Cobalt Co. Limited was still going strong. But, as a parting aside, I add that the Staffordshire Sentinel of 30 August 1915 reports that the War Office has granted permission for volunteers to obtain non-service rifles at their own expense. Some might call this "tough justice". The First World War had by then overcome all Military thinking.

A personal confession of mine concerns the difficulties of dealing with the Williamson initials and similar repeating names generation to generation. This problem is amplified by the clear mistake on the Tribute to Hugh Henshall Williamson by the 13th Company-F on a grave stone in the Church at Astbury, which I have only just recently found. The plaque has all the details relating to HHW, but the name at the top is Robert W, his father.

Hugh Henshall Williamson

Source, See Ref 13

A Collection of Personal Notes in Preparing this Book

NOTES: THE WILLIAMSON DYNASTY

(primarily developed from Mr Fords' excellent book and genealogical research on the Williamson family).

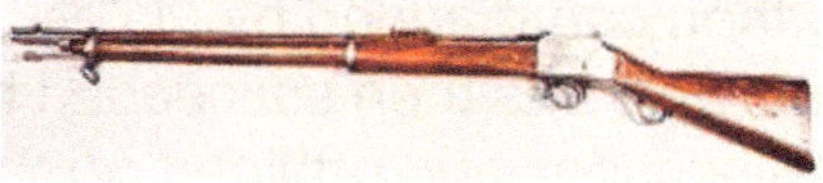

The involvement of the Williamson family in Stoke-on-Trent commences when the presumably rich landowner Robert Williamson from Mottram in Longdendale, Cheshire (1720 – 1801) invested in the pottery industry in Longport along with other coal mining interests. Pottery manufacturing in Longport was initiated by John Brindley in 1773, the younger brother of the famous Engineer James Brindley (Trent-Mersey Canal). Another potter in Longport was Edward Bourne who I believe may have been in partnership with Williamson at some point.

The Messer's Davenport were to take over and expand all these Longport pottery factories, including introduction in 1801 of the manufacture of flint-glass and ultimately stained glass. Mr Henry Davenport also purchased the Longport Hall property from Robert Williamson Snr. The other brother Mr William Davenport was elected as one of the first representatives of the Potteries Borough to Parliament (around 1835) and later took over Longport Hall.

In 1803 three companies of the early Volunteers group formed the Longport battalion, then commanded by Major Davenport, who was then succeeded by Robert Williamson, who retained command until 1814.

It was Robert Williamson's son that I refer to as Robert Williamson Jnr., (1750 - 1799), who developed even more investments in the Goldenhill and Tunstall mineral resources area and thus became a substantial coal and iron master. Robert Jnr married the widow Anne (nee Henshall) of James Brindley and begat another Robert Williamson (III) (1780 - 1869), Hugh Henshall Williamson (1783 – 1883), and John Henshall Williamson (1787 - 1856), among other siblings.

Robert Williamson Jnr., was also responsible for the building of the area commonly known as "The Barracks" in Goldenhill at a date earlier than 1839, the date of the Tithe Map, which shows a full plan of the layout of the "barracks" housing.

A further Williamson generation saw the birth of a second John Henshall Williamson (1862 – 1915). As background, the Henshall family re-Anne owned canal wharfs (on the "Grand Trunk" Canal) and transport conveyance capabilities based at Longport. It was Anne Williamson's brother Hugh Henshall that completed the workings of the Trent-Mersey Canal in 1777, following the death of James Brindley. In deference to Mr Ford's writings, we should not overlook the fact that John Gilbert (1724 – 1795) was the land agent of the Duke of Bridgewater who commissioned Brindley to build the Bridgewater and Trent-Mersey canals, the latter completed 5 years after Brindley's death.

Hugh Henshall Williamson added further investments in coal and iron ore mining, but is probably best known for his owning the Greenway Bank Estate and Greenway Hall (400 yards east of Christ Church, Tunstall). He married Ann Clowes, but had no children. Greenway Hall was earlier called Smithfield by the unfortunate potter Theophilus Smith, who came to a bitter end.

In 1848, Hugh Henshall Williamson and his brother Robert (III) founded the Goldendale Iron Company Ltd, pig iron manufacturers of Tunstall. This was one of five iron companies operating in the Potteries with its own blast furnace in 1921, but ceased operations in the 1960s.

Further, Robert Williamson III (1780 – 1869) married Ann Kinnersley the daughter of a banker from Newcastle-under-Lyme, Thomas Kinnersley, who also became a coal and iron master and who bought Clough Hall in Kidsgrove from the Gilbert family. Their offspring included (Robert) Edward Williamson, who we suggest is the Edward Williamson described in the Kidsgrove Volunteer Reserve Corps. We have identified another Edward Kennedy Williamson who is associated with Mortlake House in Congleton, but this family appears to be from a different Williamson line than our Potteries group.

Other principal land-owners in the district in the 19th century were the Sneyds (particularly Walter and Ralph) owners of Keele Hall from the early part of the 17th Century), and the Smith-Childs (owners of the former Newfield Hall).

The two generations of John Henshall Williamson mentioned basically became chemists and created and maintained the "Colour Works" in Goldenhill famed for their supplying purified cobalt minerals for input in to the glazes used in blue-coloured pottery decoration by many manufacturers, amongst a range of other chemical inputs to the pottery industry. In 1880, the first generation JHW owned the chemical works and the High Carr Ironworks (Ref 16 outlines the Memoranda Collections of the J H Williamsons held at the Archives of Swansea University and the Yorkshire Imperial Metals Records – not accessible to the author in Australia.

It seems from reviewing the Newspaper clippings of the time that the Williamson family heads or their nominees may have spent at least fifty percent of their time involved in prosecuting members of the "proletariat" in the Law Courts for failing to pay their rent, trespassing on Williamson property, petty theft of items on various properties, although I acknowledge stealing the brass fittings off a locomotive stretches that interpretation, and so on. But then, it was the Law Courts that appear to have pulled down much of the Dynasty in the course of time with the possible exception of the John Henshall Williamson segment.

Analysis of Newspaper articles also shows that the Williamson family was heavily involved in the establishment of the Voluntary military forces of the Kidsgrove/Goldenhill (even Congleton) district from the beginning of the 18th Century.

NOTES ON BANKRUPTCY IN THE WILLIAMSON FAMILY

Abstracts and Commentary on Newspaper Articles: Birmingham Daily Post, Thursday 11 November 1886; Newcastle Guardian and Silverdale, Chesterton and Audley Chronicle 11 June 1887; Birmingham Daily Post, Wednesday 14 December 1887; Congleton and Macclesfield Mercury and Cheshire General Advertiser, Saturday 18 February 1888; and Saturday 30 June, 1888; and London Evening Standard, Saturday 3 April 1886.

It is difficult to avoid using the word "evasion" to describe a number of the aspects of the proceedings of the Bankruptcy hearings of the Williamson Bros (Hugh William and William Shepherd Williamson). The proceedings on Kinnersley and Co (Edward Williamson) proved much less contentious, although the liabilities involved were over 4 times higher in magnitude.

The Williamson Brothers had the misfortune of William Shephard's passing just prior to the hearing commencement. The primary issues concerned assets previously held by Williamson Bros. It happened that a Miss Buckland had been retained as a housekeeper at Waterloo House (the home of the late William) to look after Hugh's nieces and nephews. Records showed payments to Miss Buckland, but these were "a continuation of the accounts with Hugh's nieces and nephews– some being gifts". Another property at Brown Edge had been mortgaged to a niece, Ms Rose Marie Williamson, but nothing appeared to be in the books. The examiner, Mr Tweed exclaimed "There never was such a bankrupt as this".

When the Bankrupt was asked "Did you give your niece a legal mortgage of the Brown Edge property, he replied: *So far as I suppose it is legal, I have*. When asked who prepared the mortgage (e.g., a lawyer) - the reply was *"Yes, I believe so. I can't speak off hand"*. Another Property – The Colclough property was mortgaged with the National Provincial Bank. Hugh Williamson had received rents from this property up to June that year. Further, a Craigmont Property in Wales was mortgaged to Mrs Philips, another niece, but *"it was not worth the amount specified' (£1500)*. Finally, a Property in Kent Green was mortgaged to the niece Rose Marie Williamson, and Mr Williamson had received rents from that up to September 1868, but not beyond.

There was an amount of £454 for the sale of furniture by the Bankrupt, but no records appeared in any cash book. The Bankrupt proceeded to say that *"I have been released by my old clerks to give me assistance: they have been warned not to do so".* The examiner asked "Are you making that allegation under oath?... "It is an allegation as false as an allegation can possibly be".

It appeared that Williamson Bros was indebted to Messer's Rutter and Neve to the sum of £968. Hugh Williamson replied *"They owe nothing, money is owing to them".* But when asked "How do you account for that" the answer *was "I do not know – you need to ask Mr Fox from his grave to answer that".* The examiner proceeded by asking whether Williamson Bros had borrowed £4,000 from Rutter and Neve in 1885. "Were you in difficulties at that time", Reply: *"I do not consider we were".* "Then why did you borrow that amount" Reply: *"Perhaps to pay wages".* "Do you mean it was for wages" Reply: *"I cannot tell you".* Mr Newson (the Trustee) stated that the books showed that a large proportion of this money was paid to Kinnersley and Co.

Critically, Hugh Williamson was examined at some length on the balance sheets of the

Williamson Bros holdings, The "deficiencies", as the liabilities appear to have been termed, amounted to £21,217 for the Stone Trough Colliery, £13,064 for the Goldendale operations, and £5,757 for the Goldenhill property – a total of £40,000. The Trustee noted that that figure would have been nearer £100,000 had the assets been properly entered. The Bankrupt followed by saying *"there was not a shilling put down for the outlays, recoveries and the like, which he estimated were worth more than £30,000"* – the Trustee noted that these items had been entered under "colliery accounts". When the examiner pressed issues on specific points, the bankrupt again *referred to the deceased Mr Fox*. The Registrar said that he thought the accounts as presented, compared with the evidence of the bankrupt were **most unsatisfactory.**

In a further hearing in December 1887 Mr Tweed noted that the Bankrupt (Hugh Williamson) was ordered to file within a month a further statement of accounts, and handover to the Trustee a private Cash book in use prior to 1876, and the working plans of all the collieries. He did not file the accounts until 1 week before the hearing and failed to produce the private cash book. He had handed over some "very old" colliery plans, which were of "no earthly use". The Bankrupt fenced with the questions put to him by the examiner Mr Tweed, and the Registrar threatened to adjourn the examination. The failure of Hugh William Williamson to produce documentary evidence during the bankruptcy hearings was alarming.

Kinnersley and Co

The Kinnersley and Co story proves equally intriguing, but perhaps less elusive. When Edward Williamson took possession of the Clough Hall works in 1879, the balance sheets were made up half-yearly. By 1883, Edward claimed he had put about £40,000 into the concern. At the end of 1883 the concern was hopelessly insolvent, all the capital being gone, and £50,000 of accumulated losses starting from 1881 (£16,652), 1882 (£8,142), and 1883 (£25,443). By the end of 1883 unsecured creditor's claims amounted to £38,767 of which £20,000 was due to Williamson Bros. By December 1884, a further loss of £22,400 was incurred, and the effect of the bankrupt continuing to trade increased the amount owing to Williamson Bros to £34,000. Edward Williamson was criticised for continuing to trade, but contended that *they were not insolvent and could recover within a year* in the hope of a "turn" in the market. More liabilities were incurred in 1885 – again *"one year of good trade would have paid off our debts"!*

In his first appearance at a Bankruptcy hearing Edward Williamson was informed that the total liabilities in this matter amounted to £160,077, the creditors fully secured being £38,304, and creditors of £95,296 being secured to the extent of £22,900. The liabilities thus totalling £98,982, with zero assets.

Edward was also a member of the firm Williamson Brothers (Hugh Henshall and William Sheppard), but not a managing partner. Indeed, the Williamson Brothers were amongst his main creditors. At his first bankruptcy hearing Edward was asked to outline his "current establishment". He noted that he was able to maintain his Congleton house, kept three-maid servants, two-man servants, and a boy; and had a carriage and a pair of horses. His expenditure was about £400 a year, paid for primarily from income received from his sister-in-laws.

My grievance is that here we have an example of a man continuing to live "high-off-the-hog" having been responsible for the redundancies of hundreds of hard working mine workers queuing at “soup kitchens” for sustenance. He even admitted to “he was the cause of the failure of his brothers”. As Philip Leese has communicated: “there were no signs of Williamsons in the people who donated towards the relief of the poor at that time of need; Sir Smith-Child and Captain Edward Heathcote being the main contributors” in KIdsgrove.

On another court appearance Edward was also at pains to explain that the income he received from his two sisters-in-law was paid under an arrangement between them and his departed brother William. There was no consideration for such an arrangement, which was made solely out of brotherly love. He admitted, for the second time, that he was the cause of the failure of the brothers, but in spite of that he would swear that there was no private fund from which his income was paid. There was no trust or deed, or other document by which the money was secured, but it was an entirely voluntary payment on the part of the ladies, whose separate incomes were about £1,500 and £2,000, respectively.

The matter of the Williamson Brothers “drawings” was brought up, since these separate sets of drawings were not only treated as individual assets, but the interest charged on these drawings was treated as profits. If these assets, which thus existed on paper only had been eliminated, the balance sheets would have shown that the Williamson's were insolvent years and years before their date of bankruptcy declaration.

The Examiner, Mr Tweed introduced the matter of a petition (in the matter of Ms Georgina Attwood's bankruptcy prior to her handing over the Clough Hall Works to Edward Williamson. Mr Rhoderick McKay had presented this petition in the hope that Edward Williamson might provide some kind of information, but Mr Williamson appeared to know nothing. The petition stated that the fast plant attached to the iron works, and valued at £7,800, and the loose plant valued at £1,000 and other things about the premises were to become the property of the said Mr McKay. Mr McKay said he had not received any money, and that he did not understand the order made under the petition to the High Court. Mr Edward Williamson also said he knew nothing about this matter – it came to pass that the problem rested with the National Provincial Bank, not Edward Williamson.

Incidentally, the matter of Ms Georgiana Attwood's bankruptcy was annulled in April 1886, according to a report in the London Evening Standard (3 April 1886).

NOTES: SOCIAL ISSUES

David Wood's initial discovery of the Volunteer Battalion – (1896) forms the basis of a much larger story of changes in the Social Organisation of Great Britain, and particularly for us, of the Pottery district including Goldenhill, that commenced some 37 years earlier in 1859. This story has its origins in the formation of **Rifle Volunteer Corps** (RVC) throughout Great Britain in answer to a call by the Government and the Queen (Victoria) to create a home-based military force to defend Britain from hostile threats; and in reality to provide a source of trained men to serve overseas in the event that there were losses in Britain's territories where the regular forces were defending our country's interests.

The research that I have conducted over the past months enables me to put together a fairly comprehensive story including some of the Social impacts of having Battalions of Volunteers involved in formal parades and functions, provision of music through the formation of Brass Bands, and most of all the massive growth in Shooting clubs and competitions – if you like in today's terminology 'Gun Clubs'. Add to this the social intercourse/blending that is evidenced by the numerous festivities that occurred in all the individual corps in beer houses and pubs throughout the country. Every Prize giving ceremony added to these social occasions, and the demonstration of people participating in entertaining each other through song, dance and music stands out.

If readers can identify with some of the participants involved please let me know.

NOTES: SOME PERSONAL INTERPRETATIONS

As a youngster I grew up in Sandyford in a typical working-class family, even though today some might consider me "a bit of a wild colonial boy". Nevertheless, I feel entitled to make a few observations about the Volunteer Rifle Movement that might not be considered to be that "loyal or patriotic". I think had I have lived in the 19th Century I might well have been a radical trade unionist, or at least a proponent of people's rights, but you can put that down to me being a colonial anyway.

My first observation relates to who were the people promoting the formation of the RVC Movement? It is evident that it was the wealthy, influential owners of land, mines and factories, who were in collaboration with the Authorities of the day. The scare tactics were generated in these circles which involved a high level of influence by the Church.

A similar question, who occupied the Officer ranks in these RVC's? – It was that same group of wealthy influential owners of land, mines and factories. In Kidsgrove and District

- the Williamson family. It never ceases to amaze me how the landed-gentry appear almost overnight to gain titles of senior military officers, without in many cases, any obvious military experience or military training.

The more I have read on this topic, the more I subscribe to the idea of a moderate "conspiracy" in and around the formation of the Rifle Volunteers. Such a conspiracy has many good social outcomes as well as bad, but maybe I am attracted to the bad. Recall that this period in history was also a time of agitation for the poor and their role in society, a period of "industrial unrest". Thus, the dominant argument is that if the spare time of the rebellious can be diverted by capitalising on "loyalty and patriotism" then they would be less able to rebel. Enter the Volunteer Movement - Military discipline requires training, and this takes time.

The average working-man joining the RV Corps was expected to devote not less than 2 evenings per week. after work, to military drills. In addition, they needed to be trained to shoot – so a march out of "town" would be required to a suitable shooting range, usually at least one or both of the weekend days. The up side of all this for the individual is clearly a sense of pride in "belonging", coupled with the comradery of the Company. This includes belonging to an organisation with a respectable military uniform; belonging to an organisation with a "club" building – the Drill Hall and the Armoury; belonging to a responsible group trusted with bearing arms; and belonging to a group that is respected by the community and the nation, because of what they stand for. How often did employers support their employees in these activities, either with paid time off or payment for participation?

Incidentally, my comments are not as sanguine as they may appear – it is well established in sociological history that one of the reasons for the formation of Brass Bands and Brass Band competitions was to divert the interests of the working-classes away from complaining about their lot – be it about wages, working conditions, or disruptive strikes.

Bibliography

1. John Ward, *The Borough of Stoke-upon-Trent*, Printed and published by W Lewis and Son, Finch Lane, London, 1843. (Available as a Google Book).
   A more recent book by Paul Anderton, *Called to Arms 1803 -12 in the Staffordshire-Cheshire Border Region*, Audley & District Family History Society, Stoke-on-Trent. 2016; also provides listings of some of the individual Corps members for that period, but only for Clough Hall in the context of the Potteries.

2. Ian F.W. Beckett. *Riflemen Form, A Study of the Rifle Volunteer Movement 1859 – 1908,* Pen & Sword Books Ltd, Barnsley, South Yorkshire, 1982; available on Kindle Books (no page numbers are shown, but a location number Loc out of a total number, e.g., Loc 512 of 11441. There are 11441 "Loc" in this book, so use of just the Loc number as the reference.

3. *The North Staffordshire Volunteer Regiment 1860*, summary at:

https://en.wikipedia.org/wiki/Staffordsh ire_Rangers
See also: Ray Westlake, *Tracing the Rifle Volunteers 1859 – 1908, A Guide for Military and Family Historians*, Pen & Sword Books Limited, South Yorkshire, 2010; which provides a catalogue of all Corps in all the counties for that period.

4. *Old Victorian Ordnance Survey Map (1888-1913) of Talke Pits, Staffordshire, United Kingdom*, see: https://www.archiuk.com//cgi-bin/build_nls_historic_map.pl?search_location=,%20Talke%20Pits,%20Staffordshire&latitude=53.073956&longitude=-2.255192&widgets=1

5. David Haden, *On J.R.R. Tolkien at the Butts, near Newcastle-under-Lyme, in September 1915*. See: https://potbanks.files.wordpress.com/2016/07/tolkien-paper-haden-final-rejected.pdf and

6. *The Wolstanton, Staffordshire Tithe Map 1836 -1845*, published 1839, Staffordshire Fast Track, last accessed September 2021: https://www.search.staffspasttrack.org.

uk/Details.aspx?&ResourceID=42703&PageIndex=1&KeyWord=wolstanton%20tithe&SortOrder=2

7. *Kelly's Directory of Staffordshire*, 1850 (actually Post Office Directory of Birmingham with Staffordshire and Worcester included); 1872; 1880; 1884; 1888; 1892; 1896; 1900; 1904; 1908; and 1912. Focus on Sections dealing with Military, Goldenhill and Tunstall. Prior to and including 1884 no Military listings included.

8. Newspaper Access – using the British Newspaper Archive: https://www.britishnewspaperarchive.co.uk Focus on the Birmingham Daily Post, the Staffordshire Sentinel and Commercial and General Advertiser (SSCGA), and later (1890's to 1912) the Staffordshire Sentinel (SS). The dates of articles and Newspaper source are cited in the text. (It is convenient to use the acronyms SSCGA and SS).

9. *The Army Lists for the British Army* and *Monthly Army Lists for the British Army*, 1815 through to 1936 fully digital with interactive search:

https://wiki.fibis.org/w/Army List for B ritish Army online

10. Talbot, Philip, *Accounting for Local Management Control in the Late Victorian Volunteer Army,* Journal of the Society for Army Historical Research, vol. 78, no. 314, 2000, pp. 115–123. JSTOR, *www.jstor.org/stable/44230241. Accessed 25 Aug. 2021*

11. Mow Cop Interactive History (http://www.mowcop.info/htm/industry/coal.htm)

12. *For more details on Cycling Volunteers see:* https://www.victorianvoices.net/ARTICLES/Windsor/Windsor1898B/W1898B-Cycles.pdf?fbclid=IwAR2yC VRB49AG-Uvsj2FYKdSOp27inRRg4fdTtWty9ltpx0yjmjo8qUGsaU

13. *My Family History, The History of Ramsdell Hall,* Christopher Roughan, See: *https://chrisroughan.webs.com/historyoframsdellhall.html*

14. Musson, A E. *The Great Depression in Britain 1873 – 1896; A Reappraisal.* The Journal of Economic History, Vol 19, No 2, Cambridge University Press, 1959. Pp 199 – 228. JSTOR, *www.jstor.org/stable/2114975. Accessed 25 Aug. 2021,*

15. The Archives at Swansea University, Wales have informed us of their holding a mid-19th century memoranda book of J H Williamson containing notes on metallurgical experiments, ores, processes, mineral properties etc. (ref. LAC/126/R/1). The memoranda contain information on 'Jho H Williamson, Ramsdell Hall, Cheshire' as well as the names of other family members and details of their 'dates of death', including Robert, Anne, HH, etc. (I thank Philip Leese for drawing my attention to this record).
There are also three items within the Yorkshire Imperial Metals Records relating to the Goldenhill Cobalt Company (ref. LAC/126/R):
LAC/126/R/1 - Memoranda book of J H Williamson containing notes on metallurgical experiments, ores,

processes, mineral properties etc, mid 19th century
LAC/126/R/2 - File containing miscellaneous balance sheets and correspondence papers, 1906-1909
LAC/126/R/3 - Envelope containing Agreements for the purchase of the Goldenhill Cobalt Company, 1905.

About the Author

Alan J Jones, BA(Hons), MSc, PhD, FRACI C Chem has held a number of Teaching and Research positions in universities throughout the world including the University of Keele, UK; Monash University, Australia; University of Utah, USA; Colorado State University, USA; University of Alberta, Canada; the Australian National University (ANU), Australia. He also served as a Senior Diplomat to the Federal Republic of Germany as Counsellor Industry, Science and Technology, Australian Embassy, Bonn from 1990 to 1995.

Alan held honorary appointments as Adjunct Reader, Technology and Innovation Management Centre, University of Queensland (2001 – 2003), and in a similar role at the Business School, University of Auckland, New Zealand to 2006.

Following his official "retirement" he was appointed a Visiting Fellow, National Graduate School of Management, Australian National University, working in the ARC-funded *Australian Systems of Innovation Study (2002 – 2008)*; and as an Adjunct Professor of Innovation, Management and

Performance Management, in the Division of Business, Law and Information Sciences, University of Canberra. He also served a member of the Management Committee of *Chemistry in Australia*, Royal Australian Chemical Institute (2003 – 2014).

Alan's research speciality was in multi-nuclear magnetic resonance (nmr) spectroscopy, particularly carbon-13, nitrogen-15, deuterium, and high-field proton nmr, applied to chemical and biological problems. He was Head/Senior Fellow of the National Nuclear Magnetic Resonance Centre at the ANU.

He is author/co-author of over 100 papers in refereed international scientific (but later business) journals, along with over 115 articles in more general scientific and technological publications, including interviews with distinguished Professors Carl Djerassi, Chennupati Jagadesh and Paul Anastas; and comprehensive reviews of aspects of the National Innovation Systems of nations like Finland, Taiwan, Germany that he visited for this purpose.

He is also the author of 12 monographs which cover strategic areas of science and their technological applications in industry from lasers; membranes; biomaterials; superconductivity; electrodes, sensors; titanium; and water and waste management; and even the methodology for selecting such fields of study, all published in his name by the Government of Australia.

Books by this Author

1. *Investing in Knowledge Capital: Management Imperatives*, Hardcover – January 1, Singapore Institute of Management, 2002. Shantha Liyanage and Alan J. Jones.

This book looks at the complex web of knowledge interactions by deconstructing and constructing knowledge to examine its creation, assimilation, and diffusion processes. The current views on knowledge relations and their composition are analysed to filter out those aspects which encourage economic growth.

2. *Serendipitous and Strategic Innovation – A Systems Approach to Managing Science-based Innovation,* Hardcover, 2006. Praeger Publishers. Authors include: S. Liyanage, J. Annerstedt, P Gluckman, D.Hunyor, A.J. Jones, and M. Wilson.

This work draws together the perspectives of a multidisciplinary group of professionals—medical doctors, innovation policy analysts, and academics in business management— Liyanage and his colleagues provide a thorough examination of the technology innovation process, and display its critical links with organisational functions, so that the innovation capacities of organisations can be better prepared to meet the rapid changes of our age.

3. *Australian Nobel Laureates: Adventures in Innovation*, ETN Communications, Sydney, 2004, Hardcover.

A Review of Innovation in Australia linked to the exemplars of Australia's Nobel Laureates. Alan served on the Advisory Board and contributed an article to this book.

4. *Short Stories; Some of Life's Ups and Downs,* Kindle/Amazon Books, November, 2021, Paperback, 123 pages.

A collection of Short Stories tracing events that may have brought about change in the course of their character's lives. Some of the stories are based on early childhood, some take a more adult perspective, some have elements taken from real experiences. However, most take all of these elements, but are totally fictitious. However, two of the stories about Alan, and about Heinrich, are non-fiction

Finally, Alan practices fine-art, focussing on landscapes in oils, and has won several major prizes for his work. From 2010, Alan was an active contributing member on the Executive and later President (2013 – 2018) of the Artists Society of Canberra.

Made in the USA
Monee, IL
03 November 2026

67383618R00075